POLYMERS, POLYMER BLENDS, POLYMER COMPOSITES AND FILLED POLYMERS: SYNTHESIS, PROPERTIES AND APPLICATIONS

POLYMERS, POLYMER BLENDS, POLYMER COMPOSITES AND FILLED POYMERS: SYNTHESIS, PROPERTIES AND APPLICATIONS

ABDULAKH K. MIKITAEV
MIKHAIL KH. LIGIDOV
AND
GENNADY E. ZAIKOV
EDITORS

Nova Science Publishers, Inc.
New York

For permission to use material from this book please contact us:
Telephone 631-231-7269; Fax 631-231-8175
Web Site: http://www.novapublishers.com

LIBRARY OF CONGRESS CATALOGING-IN-PUBLICATION DATA

Polymers, polymer blends, polymer composites, and filled polymers : synthesis, properties, application / Abdulakh K. Mikitaev, Mukhamed Kh. Ligidov, Gennady E. Zaikov, editors.
p. cm.
Includes index.
ISBN 1-60021-168-2
1. Polymers--Research. 2. Polymers--Industrial applications. I. Mikitaev, Abdulakh K. II. Ligidov, Mukhamed Kh. III. Zaikov, Gennadii Efremovich.
QD381.P6127 2006
620.1'92--dc22 2006010599

Published by Nova Science Publishers, Inc. ✦ New York

CONTENTS

PREFACE

"At all times countries and people are incapable of harmonic development, if their leaders live with no respect to science and scientists"
Akhmed Sevail
The Nobel Prize Laureate in Chemistry for 1999

"The only talent of mine is my maximum curiosity"
Albert Einstein

"You should have rest before getting tired and get medical treatment before getting ill"
The eastern wisdom

Polymers, polymer blends, polymer composites and filled polymers form the basis of polymer material science – the science of materials, investigation methods and control of their properties. As it is commonly known, the development of mankind passed through several important epochs. A man lived in the Stone Age, then in the Bronze Age, and later on in the Iron Age. Now we live in the Polymer Age, which is proved by some economic reasons. If we estimate the worldwide industrial production of polymers (both synthetic and natural) not by weight, but by volume, we'll get total amount of cast iron, steel, rolled stock and nonferrous metal production that reaches $400x10^6$ m^3. Hence, dynamics of the process is also important, because polymer production development is 15 – 20% more intensive than development of the metal industry. Such huge production put forward the tasks of improving quality of articles from polymers and extending the field of their application, because even a small enhancement (for instance, extension of reliable operation time of polymeric articles) appears a very important economic question.

The editors of this collection will be grateful to receive any valuable and positive comments on it, and as well as recommendations, which might be taken into account in our future works.

Prof. A.K. Mikitaev
Chairman of the Conference,
Director of the "Research Center of Composite Materials",
Moscow, Russia
Prof. M. Kh. Ligidov
Deputy Chairman, the Dean of Chemical Faculty,
Kh.M. Berbekov Kabardino-Balkarian State University,
Nal'chik, Russia
Prof. G.E. Zaikov
Deputy Chairman,
Head of Laboratory for Chemical Resistance of Polymers,
N.M. Emanuel Institute of Biochemical Physics,
Russian Academy of Sciences,
Moscow, Russia

In: Polymers, Polymer Blends, Polymer Composites… ISBN 1-60021-168-2
Eds: A.K. Mikitaev, M.K. Ligidov et al., pp.1-15

Chapter 1

POLYMER/SILICATE NANOCOMPOSITES BASED ON ORGANOMODIFIED CLAYS

A. K. Mikitaev*[1]*, O. B. Lednev*[2*]*,
***A. Yu. Bedanokov*[1] *and M. A. Mikitaev*[3]**
[1]A.N. Nesmeyanov Institute of Organoelement Compounds of RAS, 119991 Vaviliva st., 28, Moscow, Russia
[2]D.I. Mendeleev University of Chemical Technology of Russia, 125047 Miusskaya sq., 9, Moscow, Russia
[3]State Scientific Institution “Compositecenter”, 125047 Miusskii Square 9, Moscow, Russia

ABSTRACT

It should be known that a lot of studies devoted to the preparation of polymer nanocomposite materials have been investigated at resent years. The amount of such works increases intensively. The possibility of preparation such materials was shown for practically all kinds of polymerized and polycondensed polymer materials. Investigators demonstrate particular interest to the organomodified montmorillonyte as an element of nanotechnology and bearer of nanostructure with great differ between its length and thickness. In this case the organomodification is carried out by using of ionic surfactants. The using of nonionic surfactants for hydrofobization of clay’s surface also found reflection in some works. Common steady understanding is formed about investigation technique and structure of polymer nanocomposite materials and how thermo-mechanical properties depend on its structure. The increasing in amount of such investigations shows that this perspective technology will find reflection in industrial application.

Keywords: nanocomposite, organoclay, fire retardant polymer.

---

* Correspondence to: Oleg B. Lednev, D.I. Mendeleev University of Chemical technology of Russia, 125047 Miusskaya sq., 9, Moscow, Russia. mailto: lednev_oleg@mail.ru

During last years one of the most perspective fields of polymer science there is a preparation of polymer materials that have a lot of improved and new properties. Such properties can be attributed to the new kind of materials calling polymer nanocomposites that can be used in different branches of polymer applications.

To achieve improved properties in polymer composites have to use such additives as pigments, inhibitors, antioxidants, plasticizers and other compounds. Materials including the inorganic particles (oxides, nitrides, carbides, silicates etc.) are introduced to the polymer matrix in case of nanocomposites. Main our interest devoted to the polymer nanocomposite materials based on organomodified layered silicates [1]. Incompatibility of these inorganic and organic components – main problem has to be solved. There is method to overcome this problem. It's a modification of clays by organic ionic or nonionic compounds. Modified clay (organoclay) has some advantages in comparison with simple clay:

1) Organoclays can be well dispersed in polymer matrix [2].
2) Organoclays interacted with polymer's chain [3].

For preparation of such nanocomposites based on organoclays have to be used layered natural inorganic structures as montmorillonite [4, 5, 6], hectorite [3], vermiculite [7], saponin [8], kaolin, etc. Length of these layers about 220 nm, and thickness – 1nm [9, 10].

Their crystal structure consists of two fused silica tetrahedral sheets sandwiching an edge-shared octahedral sheet of either alumina or magnesia. Stacking of the layers leads to a regular Van der Waals gap between the layers, called the *interlayer* or *gallery*. Isomorphous substitution of Si4+ for Al3+ in the tetrahedral lattice, and of Al3+ for Mg2+ in the octahedral sheet generates and excess of negative charges that are normally counterbalanced by cations (Na+ or K+) residing in the interlayers (Fig.1) [11].

The organically modified clays are prepared by the addition of long chain aliphatic quarternary ammonium or phosfonium cations to sheet mineral inorganic clays. An ion exchange process is utilized to displays the inorganic cations (e.g. sodium) with organic cations, thus improving the compatibility of the organosilicate with an organic environment [2].

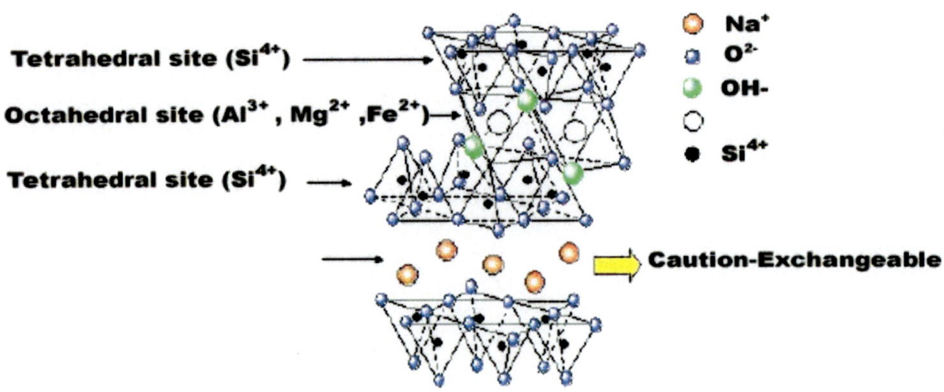

Fig. 1. Structure of layered clays

When mixed into the host polymer, exfoliation (breaking apart) of the nanophase organoclay can occur whereby the silicate sheets lose their attraction to each other. A very large increase in surface area occurs and, if the chemistry properly designed, the polymer chains can become attracted to the clay sheets. A hybrid inorganic-organic material is produced with altered properties that vary depending on the level of dispertion, the organic cation, the silicate, and the host polymer. Also can be used nonionic substances to modify clays that can be attracted to clay's surface mainly by hydrogen bonds. In some cases organoclays obtained by nonionic surfactants are more chemically stable than organoclays obtained using cation modifiers (Fig. 2) [12].

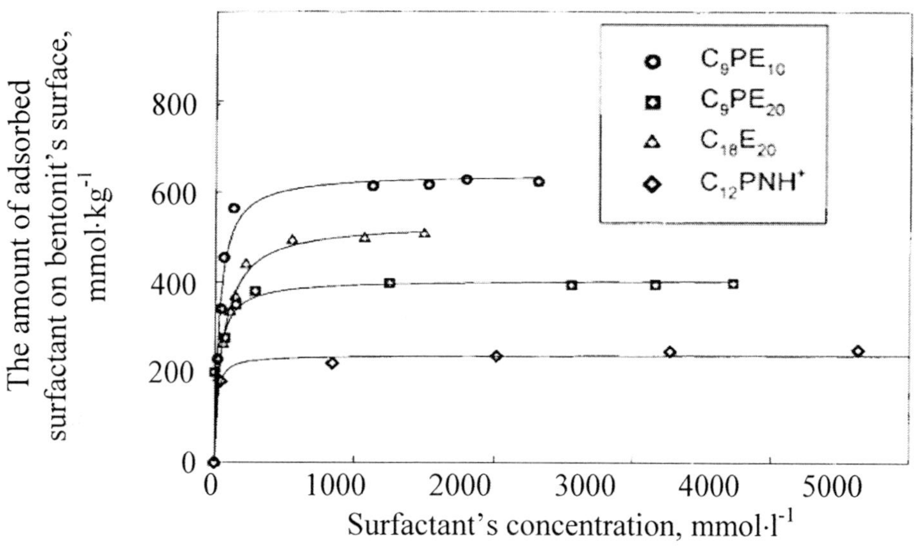

Fig. 2. I – the adsorptions of different modifiers on clay's surface

Desorbability of surfactants from bentonite depends on the mechanism of adsorption. A desorption hysteresis is generally observed when cationic surfactant is adsorbed via ion exchange (Fig. 2 II). The desorbability of surfactants from bentonite was compared by consecutively washing the organobentonite with deionized water. The nonionic surfactants in bentonite are relatively resistant tot desorption; >80% were still adsorbed after seven consecutive washes. However, for cationic surfactants, washing resulted in 25% desorption. Apparently, the cumulative effect of hydrogen bonding between individual ethylene oxide units and the bentonite surface makes organobentonite derived from nonionic surfactant chemically more stable than organobentonite derived cationic surfactant.

Their makeup is such that they can be transformed into new materials possessing the advantages of both organic materials, such as light-weight, flexibility, and good moldability, and inorganic materials, such as high strength, heat stability, and chemical resistance. The incorporation of organic/inorganic hybrids can result in materials possessing excellent of stiffness, strength and gas barrier properties with far less inorganic content than is used in conventionally filled polymer composites: the higher the degree of delamination in polymer/clay nanocomposites, the greater the enhancement of these properties [5].

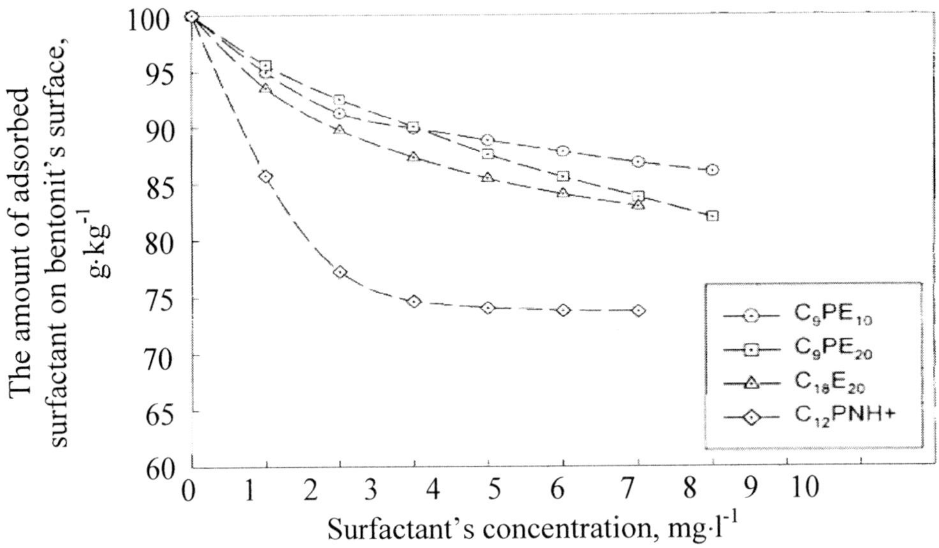

Fig. 2. II – the desorptions of different modifiers on clay's surface, where: C9PE10 – C9H19C6H4(CH2 CH2O)10OH; C9PE20 – C9H19C6H4(CH2 CH2O)20OH; C18E20 – C18H37(CH2 CH2O)20OH; C12PNH+ – C12H25C6H4NH+Cl-

At present time have been synthesized a lot of different polymer nanocomposite materials based on various kinds of polymer materials and natural inorganic fillers. The amount of such works increases very intensively (Table 1).

Table 1. Nanocomposites based on organoclay

Name of polymer	*Shorthand notation*	*Literature*
Polyacrilate	PAC_r	[13]
Polyamide	PA	[3, 14, 15]
Polybenzoksazole	PBO	[16]
Polybutylenetherephtalate	PBT	[2, 4, 18]
Polyimide	PI	[19]
Polycarbonate	PC	[20]
Polymethilmetacrilate	PMMA	[21]
Polypropylene	PP	[22, 23]
Polystyrene	PS	[21]
Polysulfone	PS_n	[24]
Polyurethane	PU	[25]
Polyethyleneterephtalate	PET	[26]
Polyethylene	PE	[27]
Epoxy	EP	[28]

Several methods have been used to obtain polymer nanocomposites by using organoclays [29-32], i.e. solution intercalation [33-39], melt intercalation [40, 41], and in situ interlayer intercalation [30, 42, 43]. Among them, in situ interlayer polymerization relies on swelling of the organoclay by the monomer, followed by in situ polymerization initiated thermally or by the addition of a suitable compound. The chain growth in the clay galleries accelerates clay exfoliation and nanocomposite formation.

This technique of in situ interlayer polymerization is also particularly attractive due to its versatility and compatibility with reactive monomers and is beginning to be used for commercial applications. However, there is ample evidence that nanocomposites can also be formed by melt processing in extruders. There are many reasons why melt processing may be more preferred method for producing nanocomposites for commercial use. Additionally, other approaches, such as the sol–gel process [44, 47] and monomer/polymer grafting to clay layers, have resulted in organic/inorganic polymer hybrids.

In the process of melt intercalation, the layered silicate is mixed with a molten polymer matrix. If the silicate surfaces are sufficiently compatible with the chosen polymer, then the polymer can enter the interlayer space and form an intercalated or an exfoliated nanocomposite. Otherwise, in situ intercalation polymerization is a method based on the use of one or more monomers that may be in situ linearly polymerized or cross linked and was the first method used to synthesize polymer-layered silicate nanocomposites based on polyamide 6. The in situ intercalation method relies on the swelling of the organoclay due to by the monomer, followed by in situ polymerization initiated thermally or by the addition of a suitable compound. The chain growth in the clay galleries triggers clay exfoliation and nanocomposite formation. Thus, an advantage of the in situ method is the preparation of polymer hybrids without physical or chemical interactions between the organic polymer and the inorganic material.

According to the early work of Giannelis [49], in general two types of hybrid structures can be obtained upon PCN preparation: *intercalated*, in which a single, extended polymer chain is intercalated between the silicate layers, resulting in a well-ordered multilayer with alternating polymer/inorganic host layers and a repeat distance of a few nanometers, and *disordered* or *delaminated*, in which the silicate layers (1 nm thick) are exfoliated and dispersed in a continuous polymer matrix (Fig 3). The best performances are commonly observed for the exfoliated nanocomposites; the two situations can, however, coexist in the same material. In any case, to make a successful nanocomposite it is very important to be able to disperse the inorganic material throughout the polymer. If a uniform dispersion is not achieved, agglomerates of inorganic materials are found within the host polymer matrix, thus limiting improvement.

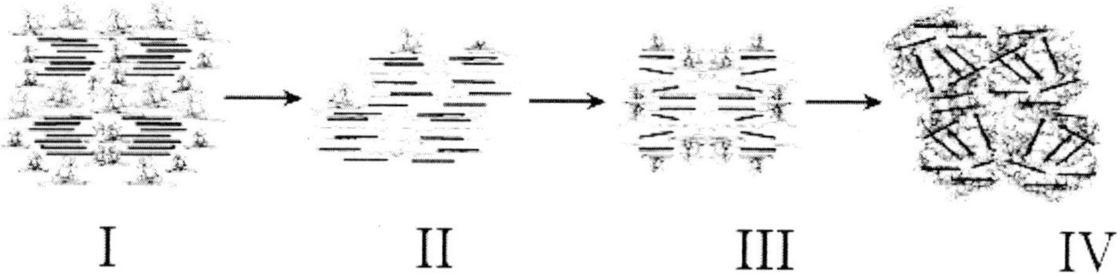

Fig. 3. Schematic illustration of nanocomposite formation

Unseparated MMT layers, after introduction into the polymer, are often referred to as tactoids. The term intercalated describes the case where a small amount of polymer moves into the gallery spacing between clay platelets, but causes less than 2-3 nm separation between the platelets. Exfoliation or delamination occurs when polymer further separates the clay platelets e.g. by 8-10 nm or more. A well-delaminated and dispersed nanocomposite consists of delaminated platelets distributed homogeneously in the polymer [4, 5, 49, and 50].

One of the methods to study the dispertion of organoclay in nanocomposite is wide-angle XRD-diffraction. Thus, on figure 4 are shown XRD-data obtained for Na^+MMT, modified clay and PBT-fibers with different loading of organoclay.

The characteristic peak for pristine clay, Na^+MMT, appears at $2\theta=8.56°$ (d=1.2 nm). For Na^+MMT reacted with alkyl amine, NCT-MMT, this peak is broadened and shifted to $2\theta=3.98°$ (d=2.6 nm), suggesting that the clay is swollen to the range of the d spacing. In general, a greater interlayer spacing should be advantageous in the intercalation of polymer chains. It should also lead to easier dissociation of the clay, which should result in hybrids with better dispersions of clay. In addition to the main diffraction peak, an additional small peak is observed at $2\theta=7.13°$ (d=1.43 nm). This secondary peak may be related to the XRD spectra of the organoclay itself. Fig.3 also shows the X-ray diffraction (XRD) curves of pure PBT and of PBT hybrid fibers with 2–5 wt% organoclay loadings. Pure PBT synthesized with an MMT interlayer exhibits its usual XRD peaks. However, in the cases of the 2 and the 3 wt% PBT hybrids, the curves show no characteristic organoclay peaks in the range of the $2\theta=2$–$8°$; that is, the peak corresponding to the basal spacing has disappeared. In the cases of the PBT hybrids with 4 and 5 wt% organoclay loadings, however, a small peak is observed at $2\theta=5.44°$ (d=1.88 nm). This indicates that agglomeration of a small part of the clay has occurred in the PBT matrix [4].

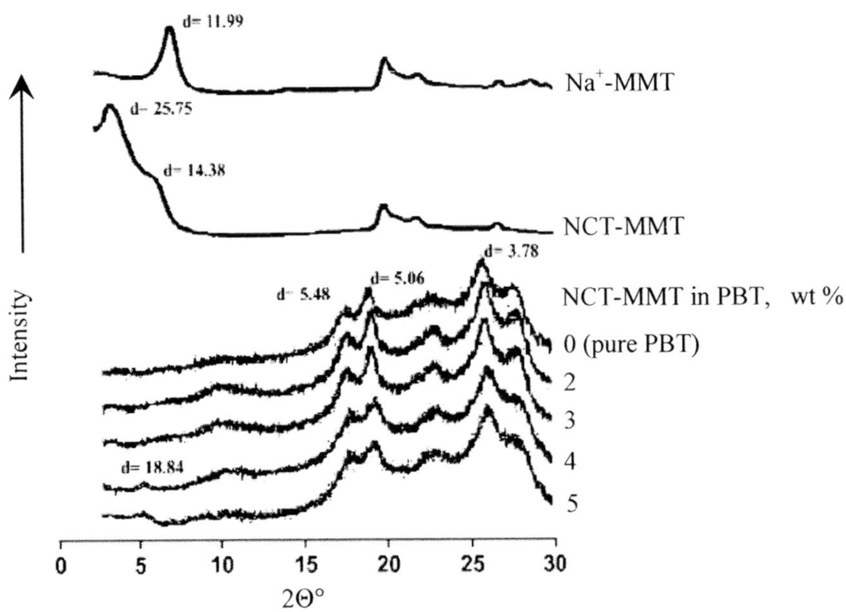

Fig. 4. XRD patterns for clay, organoclay, and PBT hybrid fibers with various organoclay contents

XRD-data also were obtained for other polymers (Fig. 5) [4, 49, and 50]. XRD is most useful for the measurement of the d-spacing of ordered immiscible and ordered intercalated polymer nanocomposites with clay, but it may be insufficient for the measurement of disordered and exfoliated materials that give no peak. The organoclay dispersion in also has to be crosschecked further by using the SEM and TEM data [53, 54].

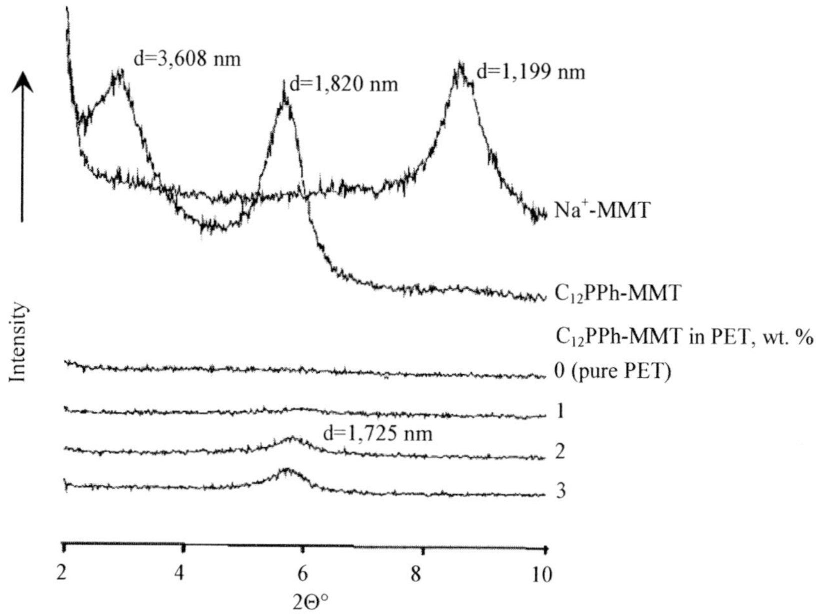

Fig. 5. XRD-data for clay, organoclay and nanocomposite PET/organoclay

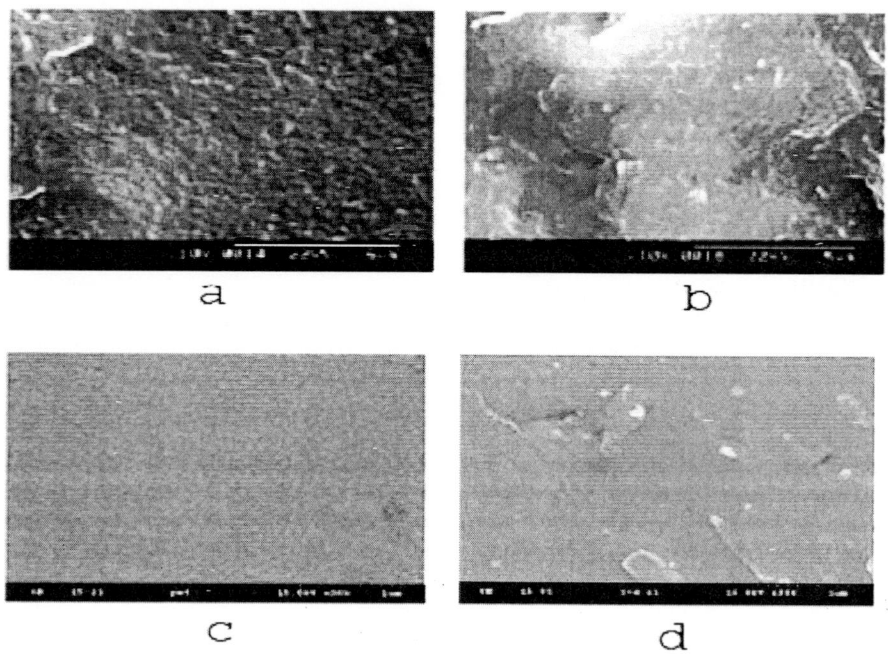

Fig. 6. SEM photomicrographs of (a) 0% (pure PBT); (b) 3% organic-MMT in PBT hybrid fibers; (c) 0% (pure PET); (d) 3% organic-MMT in PET hybrid fibers

SEM micrographs of the fractured surfaces of PBT hybrid fibers prepared with different clay contents are compared in Fig. 6, 7. The micrographs of the pure PBT/PET (Fig. 6(a, c)) and the PBT/PET hybrid fiber containing 3 wt% organoclay (Fig. 6(b, d)) show smooth surfaces due to better dispersed clay particles.

Conversely, fig. 7(a, b, c, d) show voids and some deformed regions that may result from the coarseness of the fractured surface. However, the fractured surfaces were more deformed when higher contents of organoclay were used in the hybrids. This is probably a consequence of the agglomeration of clay particles [55, 56].

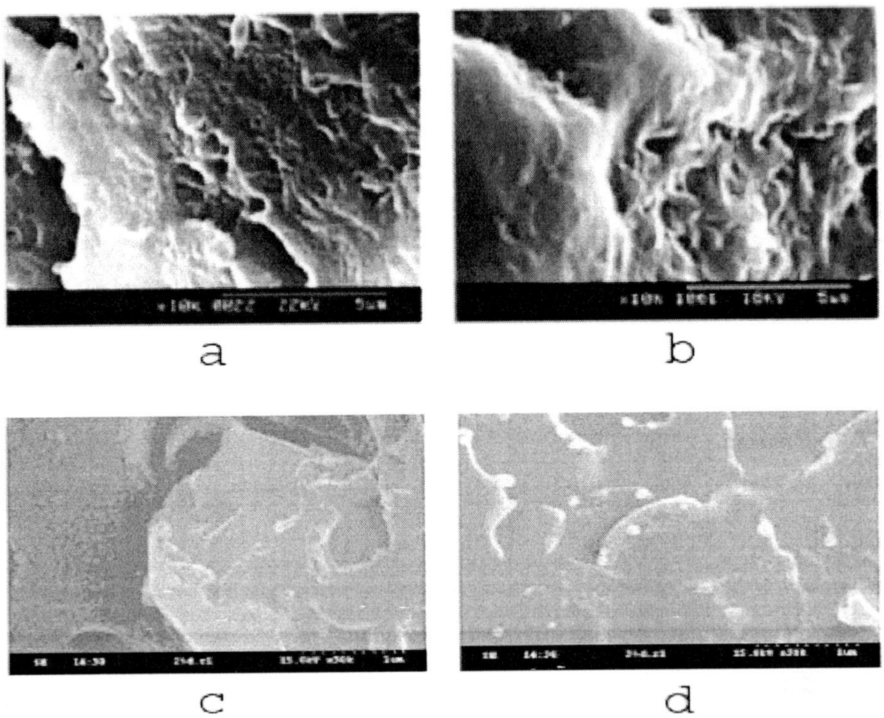

Fig. 7. SEM photomicrographs of (a) 4% (pure PBT); (b) 5% organic-MMT in PBT hybrid fibers; (c) 4% (pure PET); (d) 5% organic-MMT in PET hybrid fibers

More direct evidence for the formation of a true nano-scaled composite was provided by TEM analysis of an ultramicrotomed section. The TEM micrographs are presented in Fig. 8, 9. The dark lines are the intersections of 1 nm-thick clay sheets, and the spaces between the dark lines are the interlayer spaces. Some of the clay layers of Fig. 6 show individual dispersion of delaminated sheets in the matrix, as well as regions where the regular stacking arrangement is maintained with a layer of polymer between the sheets. Although a face-to-face layer morphology is retained, the layers are irregularly separated by, 4–10 nm of polymer.

For the 4 and the 5 wt% organoclay-loaded PBT/PET hybrid fibers (Fig. 9) however, some of the clay is well dispersed in the PBT/PET matrix, and some of it is agglomerated to a size of approximately 4-8 nm. This is consistent with the XRD results shown in Fig. 4, 5.

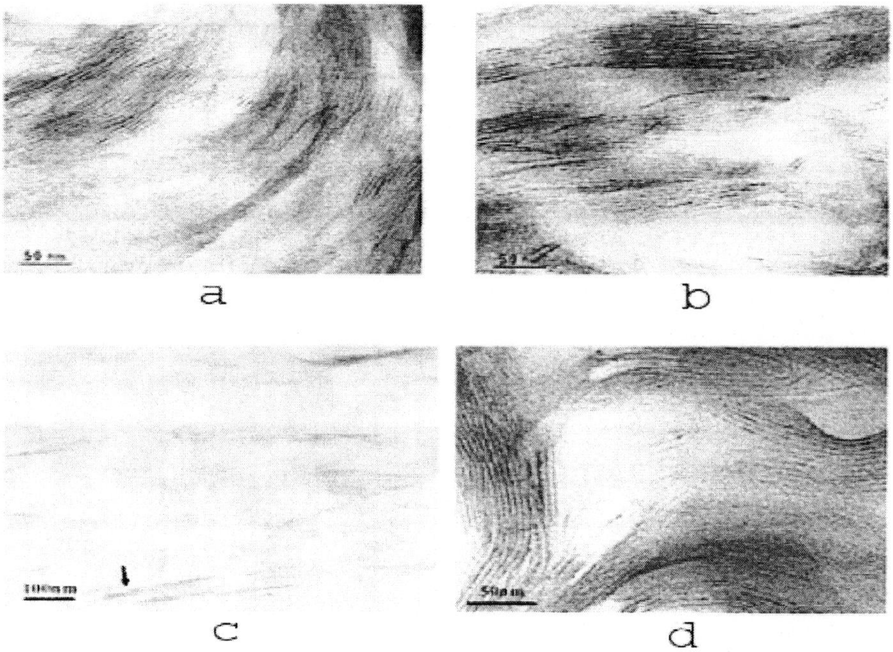

Fig. 8. TEM photomicrographs of (a) 2% organoclay in PBT; (b) 3% organoclay in PBT; (c) 1% organoclay in PET; (d) 1% organoclay in PET

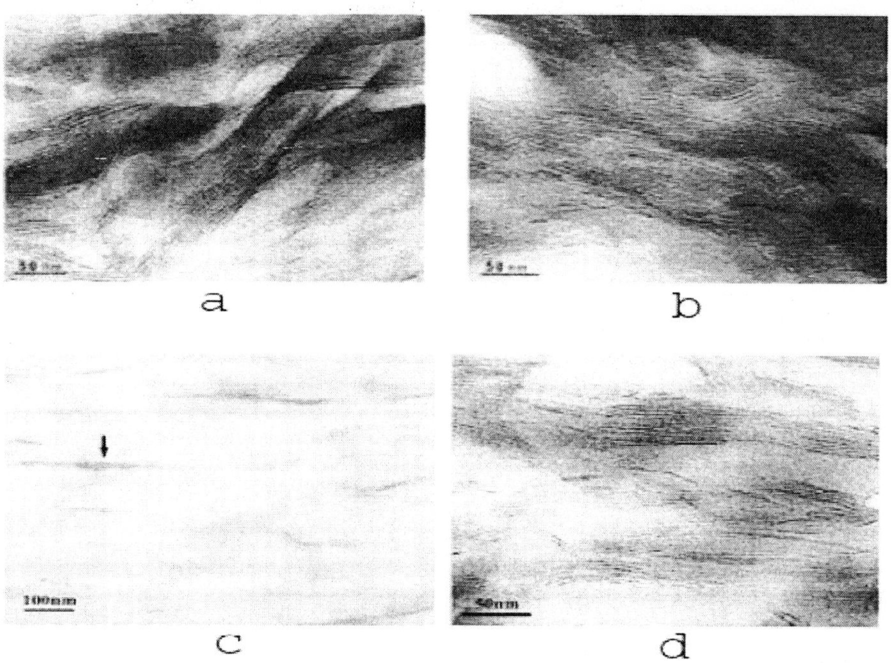

Fig. 9. TEM photomicrographs of (a) 4% organoclay in PBT; (b) 5% organoclay in PBT; (c) 3% organoclay in PET; (d) 3% organoclay in PET

From the results of XRD and electron micrographs, the morphology at a low organoclay content (<3 wt %) presents a mixture of intercalated and partially exfoliated features. The dispersion is better at a lower organoclay loading than at a high organoclay loading.

Using of organoclays as nanoparticles in polymer materials can change such properties as glass transition temperature (T_g), melting temperature (T_m) and decomposition temperature (T_D), ultimate tensile strength (Ult. Str.), initial modulus (Ini. Mod.), percent of elongation at the break (E.B.), gas-, moisture barrier properties.

The thermal and mechanical properties of PBT and PET (polyethylene terephthalate) hybrids with different contents of organoclay are listed in Table 2.

The glass transition temperatures (Tg) of the PBT hybrids increased from 27 to 33 °C with increasing clay loading from 0 to 2 wt% and then remained fairly constant up to 5 wt% organoclay. The increase in the Tg of these hybrids may be the result of two factors [57-60]. First, the effect of a small amount of dispersion of the clays on the free volume of PBT is significant and has an influence on the glass transition temperature of the PBT hybrids. Second, confinement of intercalated polymer chains within the clay galleries prevents segmental motions of the polymer chains. The endothermic peak of pure PBT appears at 222 °C and corresponds to the melting temperature (Tm). Similar to the result for Tg; the DSC thermograms show that the value of Tm increases from 222 to 230 °C with increasing organoclay content up to 2 wt%, and then remained constant for additional organoclay loading up to 5 wt% (Table 2). This increase in the thermal behavior of the hybrids may result from the heat insulation effect of the clay layer structure, as well as from the interaction between the organoclay and PBT molecular chains [59, 60].

Table 2. Main properties of polymer nanocomposites

Properties	Composite structure								
	Polybutylene terephthalate + NCT-montmorillionite					Polyethylene terephthalate + C_{12}PPh-montmorillonite			
	Organoclay quantity, %								
	0	2	3	4	5	0	1	2	3
Viscosity, Dl/g	0,84	1,16	0,77	0,88	0,86	1,02	1,26	0,98	1,23
T_g, °C	27	33	34	33	33	---	---	---	---
T_m, °C	222	230	230	229	231	245	247	245	246
T_d, °C	371	390	388	390	389	370	375	384	386
W_{tR}^{600c}, %	1	6	7	7	9	1	8	15	21
Ultimate strength, MPa	41	50	60	53	49	46	58	68	71
Initial modulus, GPa	1,37	1,66	1,76	1,80	1,86	2,21	2,88	3,31	4,10
Percent of elongation at break, %	5	7	6	7	7	3	3	3	3

Thermal stability of nanocomposites based on PBT/PET studied by thermo gravimetric analysis (TGA) is shown in table 2 and figures 10, 11 [4, 5].

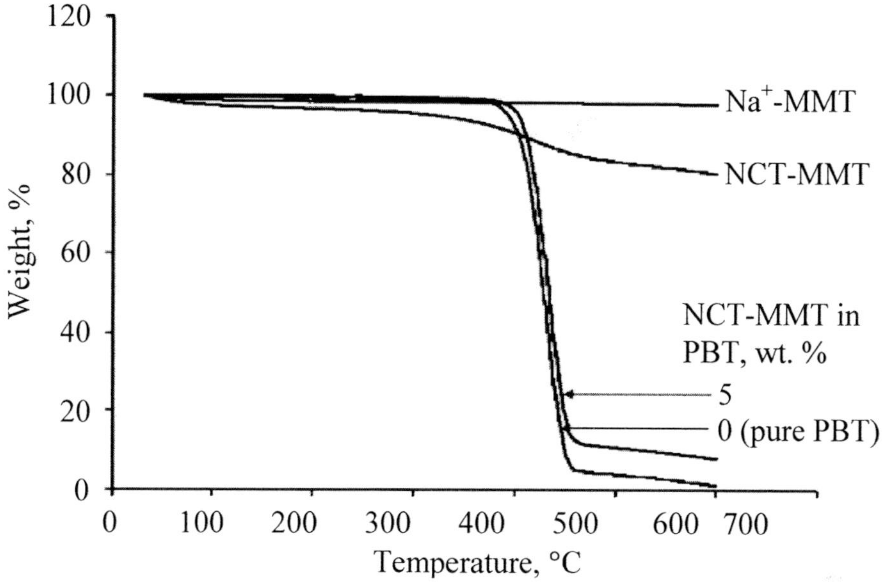

Fig. 10. TGA curves of montmorillonites, PBT and nanocomposite PBT/organoclay

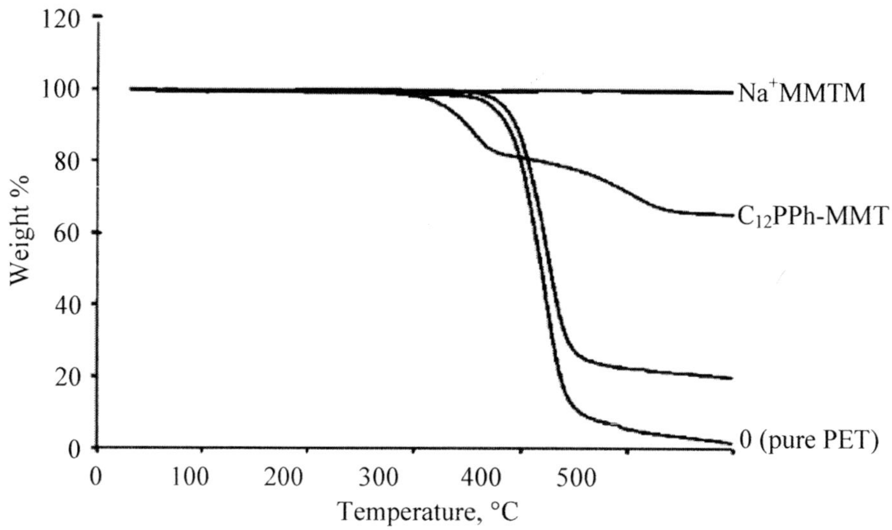

Fig. 11. TGA curves of montmorillonites, PET and nanocomposite PET/organoclay

The weight loss due to the decomposition of PBT and its hybrids was nearly the same until a temperature of 350°C (Fig. 10). Above 350°C, $T°_D$ was influenced by the organoclay loading in the hybrids. The addition of clay enhanced the performance by acting as a superior insulator and as a mass-transport barrier to the volatile products generated during decomposition [63, 64, and 65]. This kind of improvement in thermal stability has also been observed in many systems of hybrids [66, 67]. The weight of the residue at 600°C increased with increasing clay loading from 0 to 5%, ranging from 1 to 9%. This enhancement of the char formation is ascribed to the high heat resistance due to the clay itself. Considering the above results, we find it consistently believable that the introduction of an inorganic clay component into an organic polymer can improve polymer's thermal properties due to the good thermal stability of the clay.

As it shown in table 2 at DR (draw rate) = 1, the ultimate tensile strength of the PBT/NCTMMT hybrid fibers increases with the addition of clay up to a critical clay loading, and then decreases above that critical content. For example, the strength of 3 wt% PBT hybrid fibers is 60 MPa, which is about 50% higher than that of pure PBT (41 MPa). When the amount of organoclay in PBT reaches to 5 wt%, the strength has decreased again to 49 MPa. This suggests that the NCT-MMT domains are more agglomerated above 3 wt% organoclay content in the PBT matrix [61, 68, and 69].

However, the initial modulus monotonically increased with increasing organoclay content in the PBT (Table 2). The value of the initial modulus increased constantly from 1.37 to 1.86 GPa with increasing NCTMMT content up to 5 wt%. This enhancement modulus is ascribed to the high resistance exerted clay itself. Additionally, the stretching resistance and oriented backbone of the polymer chain contributes to the enhancement of the modulus. Percent elongations at the break of all hybrids were 7%. These values remained constant for organoclay loadings from 2 to 5 wt%.

As Table 2 shows, the values of the strength and the initial modulus for pure PBT fibers were enhanced with increasing DR. However, as expected for the case of a flexible coil polymer, the increases in the tensile strength and the modulus with increasing DR were insignificant for pure PBT. Table 3 shows that for pure PBT, the strength and the modulus values increased. On the other hand, the values of the strength and the initial modulus of the hybrid fibers decreased with increasing DR. For the hybrid with 3 wt% organoclay, for example, when the DR was increased from 1 to 6, the ultimate strength and the initial modulus decreased. These trends with increasing DR were observed in all systems containing 2–5 wt% organoclay. This declination in the tensile properties seems to be the result of debonding between the organoclay and the PBT and the presence of many micro-sized voids due to excess stretching of the fibers [71-73].

Table 3. Effect of the DR on the tensile properties of PBT nanocomposite fibers

Organoclay loading, %	Ultimate strength, MPa			Initial modulus, GPa		
	DR=1	DR=3	DR=6	DR=1	DR=3	DR=6
0 (pure PBT)	41	50	52	1,37	1,49	1,52
3	60	35	29	1,76	1,46	1,39

Finally it has to be noted that the amount of studies devoted to the polymer nanocomposite materials increases very intensively. It was shown the ability of using of practically all kinds of polymer materials. Common understanding about investigation methods and structure of nanocomposite polymer materials, about relationships between properties and specific structure of nanocomposites is formed. Increasing of studies in this field proves that it will become very perspective approach to prepare in industrial scales new polymer materials with a lot of improved properties.

REFERENCES

[1] Kaladzhyan, A.A., Lednev, O.B., Mikitaev, M.A. Nanocomposite polymer materials based on organoclay // *Thesis of report on International Conference* Dedicated to 50th Anniversary of A.N. Nesmeyanov Institute of Organoelement Compounds (INEOS) Russian Academy of Sciences. Moscow, 2004, P31.

[2] Delozier, D.M., Orwoll, R.A., Cahoon, J.F., Ladislaw, J.S., Smith, J.G., Connell, J.W.. *Polymer*, 2003;44:2231-2241.

[3] Delozier, D.M., Orwoll, R.A., Cahoon, J.F., Johnston, N.J., Smith, J.G., Connell, J.W.. *Polymer*, 2002;43:813-822.

[4] Chang, J.-H., An, Y.U., Kim, S.J., *S. Im. Polymer*, 2003;44:5655-5661.

[5] Chang, J.-H., Kim, S.J., Joo, Y.L., *S. Im. Polymer*, 2004;45:919-926.

[6] Mikitaev, A.K., Kaladzhyan, A.A., Lednev, O.B., Mikitaev, M.A. Nanocomposite polymer materials based on organoclays.// *Electronic journal "Investigated in Russia"*, 83, p. 912-922, 2004. http://zhurnal.ape.relarn.ru/articles/2004/083.pdf

[7] Kelly, P., Akelah, A., Moet, A. *J. Mater. Sci.* 1994. V.29. P.2274—2280.

[8] Chang, J.-H., An, Y.U., Cho, D., Giannelis, E.P.. *Polymer*, 2003;44:3715-3720.

[9] Yano, K, Usuki, A, Okada, A. *J Polym Sci, Part A*: *Polym Chem* 1997;35:2289.

[10] Garcia-Martinez, JM, Laguna, O, Areso, S, Collar, EP. *J Polym Sci, Part B: Polym Phys* 2000;38:1564.

[11] Pinnavaia, TJ, Beall, GW, editors. *Polymer–Clay Nanocomposites*. New York: Wiley; 2000.

[12] Shen, Y.-H.. *Chemosphere*, 2001;44:989-995

[13] Chen, Z, Huang, C, Liu, S, Zhang, Y, Gong, K. *J Apply Polym Sci* 2000;75:796-801.

[14] Okado, A, Kawasumi, M, Kojima, Y, Kurauchi, T, Kamigato, O. *Mater Res Soc Symp Proc* 1990;171:45.

[15] Leszek, A. Utracki, Jorgen Lyngaae-Jorgensen. *Rheologica Acta*, 2002;41: 394-407

[16] Hsu, S.L.-C., Chang, K.-C. *Polymer*, 2002;43:4097-4101.

[17] Wagener, R., Reisinger, T.J.G. *Polymer*, 2003;44:7513-7518.

[18] Li, X, Kang, T, Cho, WJ, Lee, JK, Ha, CS. *Macromol Rapid Commun.*

[19] Tyan, H.-L., Liu, Y.-C., Wei, K.-H. *Polymer*, 1999;40:4877-4886.

[20] Vaia, R, Huang, X, Lewis, S, Brittain, W. *Macromolecules* 2000;33:2000-4.

[21] Okamoto, M, Morita, S, Taguchi, H, Kim, Y, Kotaka, T, Tateyama, H. *Polymer* 2000;41:3887-90.

[22] Chow, W.S., Mohd Ishak, Z.A., Karger-Kocsis, J., Apostolov, A.A., Ishiaku, U.S. *Polymer*, 2003;44:7427-7440.

[23] Antipov, E.M., Huseva, M.A., Herasin, V.A., Korolyov, Yu.M., Rebrov, A.V., Fisher, H.R., Razumovskaya, I.V. Structure and deformation behavior of nanocomposites based on polypropylene and modified clays. // *Highmolecular compounds*. A. 2003. V.45. №11. P. 1885-1899.

[24] Sur, G, Sun, H, Lyu, S, Mark, J. *Polymer* 2001;42:9783-9.

[25] Wang, Z, Pinnavaia, T. *Chem Mater* 1998;10:3769-71.

[26] Chang, J.-H., Kim, S.J., Joo, Y.L., *S. Im. Polymer*, 2004;45:919-926.

[27] Antipov, E.M., Huseva, M.A., Herasin, V.A., Korolyov, Yu.M., Rebrov, A.V., Fisher, H.R., Razumovskaya, I.V. Structure and deformation behavior of nanocomposites

based on polyethylene and modified clays. // *Highmolecular compounds*. A. 2003. V.45. №11. P. 1874-1884.

[28] Lan, T, Kaviartna, P, Pinnavaia, T. *Proceedings of the ACS PMSE* 1994;71:527-8.

[29] Pinnavaia, TJ. *Science* 1983;220:365.

[30] Messersmith, PB, Giannelis, EP. *Chem Mater* 1993;5:1064.

[31] Giannelis, EP. *Adv Mater* 1996;8:29.

[32] Gilman, JW. *Appl Clay Sci* 1999;15:31.

[33] Greenland, DG. *J Colloid Sci* 1963;18:647.

[34] Chang, JH, Park, KM. *Polym Engng Sci* 2001;41:2226.

[35] Greenland, DG. *J Colloid Sci* 1963;18:647.

[36] Chang, JH, Park, KM. *Polym Eng Sci* 2001;41:2226.

[37] Vaia, RA, Ishii, H, Giannelis, EP. *Adv Mater* 1996;8:29.

[38] Vaia, RA, Jandt, KD, Kramer, EJ, Giannelis, EP. *Macromolecules* 1995;28:8080.

[39] Fukushima, Y, Okada, A, Kawasumi, M, Kurauchi, T, Kamigaito, O. *Clay Miner* 1988;23:27.

[40] Vaia, RA, Ishii, H, Giannelis, EP. *Adv Mater* 1996;8:29.

[41] Vaia, RA, Jandt, KD, Kramer, EJ, Giannelis, EP. *Macromolecules* 1995;28:8080.

[42] Fukushima, Y, Okada, A, Kawasumi, M, Kurauchi, T, Kamigaito, O. *Clay Miner* 1988;23:27.

[43] Akelah, A, Moet, A. *J Mater Sci* 1996;31:3589.

[44] Chvalun, S.N. *Nature* 2000, №7.

[45] Brinker, C.J., Scherer, G.W. *Sol-Gel Science.* Boston, 1990.

[46] Mascia, L, Tang, T. *Polymer* 1998;39:3045.

[47] Tamaki, R, Chujo, Y. *Chem Mater* 1999;11:1719.

[48] Dennis, H.R., Hunter, D.L., Chang, D., Kim, S., White, J.L., Cho, J.W., Paul, D.R. *Polymer,* 2001;42:9513-9522.

[49] Giannelis, E.P., *Adv. Mater.* 8 (1996) 29–35.

[50] Kornmann, X., Lindberg, H., Berglund, L.A., *Polymer* 42 (2001) 1303–1310.

[51] Voulgaris, D., Petridis, D.. *Polymer*, 2002;43:2213-2218.

[52] Tyan, H.-L., Liu, Y.-C., Wei, K.-H.. *Polymer*, 1999;40:4877-4886.

[53] Davis, CH, Mathias, LJ, Gilman, JW, Schiraldi, DA, Shields, JR, Trulove, P, Sutto, TE, Delong, HC. *J Polym Sci, Part B: Polym Phys* 2002;40:2661.

[54] Morgan, AB, Gilman, JW. *J Appl Polym Sci* 2003;87:1329.

[55] Chang, JH, An, YU, Sur, GS. *J Polym Sci Part B: Polym Phys* 2003;41:94.

[56] Chang, JH, Park, DK, Ihn, KJ. *J Appl Polym Sci* 2002;84:2294.

[57] Xu, H, Kuo, SW, Lee, JS, Chang, FC. *Macromolecules* 2002;35:8788.

[58] Haddad, TS, Lichtenhan, JD. *Macromolecules* 1996;29:7302.

[59] Mather, PT, Jeon, HG, Romo-Uribe, A, Haddad, TS, Lichtenhan, JD. *Macromolecules* 1996;29:7302.

[60] Hsu, SLC, Chang, KC. *Polymer* 2002;43:4097.

[61] Chang, JH, Seo, BS, Hwang, DH. *Polymer* 2002;43:2969.

[62] Fornes, TD, Yoon, PJ, Hunter, DL, Keskkula, H, Paul, DR. *Polymer* 2002;43:5915.

[63] Chang, JH, Seo, BS, Hwang, DH. *Polymer* 2002;43:2969.

[64] Fornes, TD, Yoon, PJ, Hunter, DL, Keskkula, H, Paul, DR. *Polymer* 2002;43:5915.

[65] Mikitaev, A.K., Kaladzhyan, A.A., Lednev, O.B., Mikitaev, M.A., Davydov, E.M. Nanocomposite polymer materials based on organoclays with enhanced fire resistance.

// *Electronic journal "Investigated in Russia",* 129, p. 1365-1370, 2004. http://zhurnal.ape.relarn.ru/articles /2004/129.pdf

[66] Wen, J, Wikes, GL. *Chem Mater* 1996;8:1667.

[67] Zhu, ZK, Yang, Y, Yin, J, Wang, X, Ke, Y, Qi, Z. *J Appl Polym Sci* 1999;3:2063.

[68] Lan, T, Pinnavaia, TJ. *Chem Mater* 1994;6:2216.

[69] Masenelli-Varlot, K, Reynaud, E, Vigier, G, Varlet, J. *J Polym Sci Part B: Polym Phys* 2002;40:272.

[70] Yano, K, Usuki, A, Okada, A. *J Polym Sci Part A: Polym Chem* 1997;35:2289.

[71] Shia, D, Hui, Y, Burnside, SD, Giannelis, EP. *Polym Engng Sci* 1987;27:887.

[72] Curtin, WA. *J Am Ceram Soc* 1991;74:2837.

[73] Chawla, KK. *Composite materials science and engineering*. New York: Springer; 1987.

In: Polymers, Polymer Blends, Polymer Composites... ISBN 1-60021-168-2
Eds: A.K. Mikitaev, M.K. Ligidov et al., pp.17-24

Chapter 2

STRUCTURE AND PROPERTIES OF COMPOSITIONS ON THE BASIS OF MIXES OF EPOXYNOVOLAIC AND PHENOLFORMALDEHYDE PITCHES

Mikhail Kh. Ligidov*
Kabardino-Balkarian State University, Nalchik

ABSTRACT

The results of studying laws of processes of structure – formation in the diluted and concentrated solutions of mixes of epoxynovolaic (EV) and phenolformaldehyde (PF) pitches in various diluents environments and in films are presented in this work. The adjectives are to choose an optimal pairs of compounds and to select the time - temperature conditions of producing a wearproof working layer of hard magnetic disks.

Keywords: epoxynovolaic pitches, phenolformaldehyde pitches, characteristic viscosity

INTRODUCTION

Materials on a basis of epoxynovolaic pitches possess a unique complex of technological and operational properties. High adhesion to many metals, small shrinkage in process of solidification, chemical stability and high durability – all this provides their successful use as binding working layers of magnetic recording medium [1].

Studying laws of formation of mesh structure and properties of the unfilled and filled mixes of epoxynovolaic and phenolformaldehyde pitches during their solidification allows to establish the mechanism of solidification process of reactionary mix and to offer optimal time

* Correspondence to: Kabardino-Balkarian State University, Nal'chik, Russia, Chernyshevski 173, 360004, mailto: deanchem@ns.kbsu.ru

– temperature modes of producing a working layer of hard magnetic disks with high operational characteristics.

The results of studying laws of processes of structure – formation in the diluted and concentrated solutions of mixes of epoxynovolaic (EV) and phenolformaldehyde (PF) pitches in various diluents environments and in films are presented in this work. The adjectives are to choose an optimal pairs of compounds and to select the time - temperature conditions of producing a wearproof working layer of hard magnetic disks.

THEORY

On the basis of changing values of characteristic viscosity [η] (which are associated with the rotation and elastic – viscous deformation of macromolecular balls in a stream of solvent) and Hagging's constant one may determine the thermodynamic affinity of solvent to studied polymers or their mixes [2,3]. [η] is a measure of additional losses of energy during spreading of the solution). Hagging's constant depends on a degree of interaction of polymeric molecules with solvent and is determined by the formula

$$k' = \frac{\eta_{cp} - [\eta]}{[\eta]^2 \cdot c} \tag{1}$$

Negative deviations from additive values of characteristic viscosity [η] witnessing bad overlapping of EV and PF molecules for solutions of these polymers in mixes of the solvents containing etiltselozolv, isoforon, cyclohexanon and dimethylformamide in various parities are found. Values of Hagging's constant k' appear to have positive deviation from the additivity for mentioned solutions testifying that energy of interaction of molecules of the dissolved polymers with molecules of diluent medium exceeds energy of interaction polymer 1 – polymer 2. This indicates that interaction between molecules of solvent and molecules of both epoxynovolaic and phenolformaldehyde pitch is more preferable than interactions between molecules of the pitches dissolved in such solutions, i.e. speaks about bad overlapping of polymers in these solvents.

The best overlapping of epoxynovolaic pitches with phenoformaldehyde was attained in diluent medium containing isoforon and etiltselozolv in the ratio 2:1. As is seen from fig. 1, change of value [η] in solutions of EV:PF mixes obeys the additive law for all structures:

$$[\eta] = [\eta_1] \cdot W_1 + [\eta]_2 \cdot W_2, \tag{2}$$

where W_1 and W_2 – weight fractures of pitch compounds; $[\eta]_1$ and $[\eta]_2$ – characteristic viscosities of components.

The Hagging's constant k' changes with negative deviation from additivity for these solutions. That points good solubility of EV and PF in each other without formation of aggregated particles.

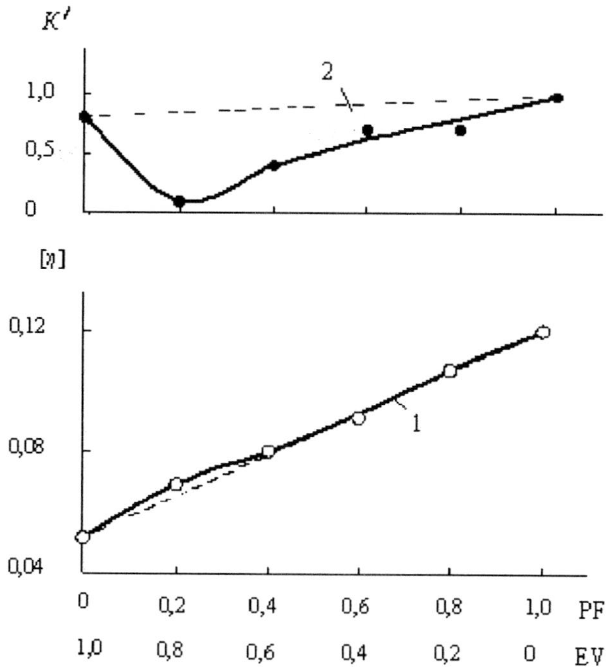

Figure 1. Dependencies of [η] (1) and *K'* (2) via composition of mix of EV and PF pitches in diluent medium containing isoforon and etiltselozolv in ratio 2:1

Basing mentioned we recommend diluent mix which contains isoforon and etilcelozolv in ratio 2:1 to obtain the solution of epoxynovolaic and phenolformaldehyde pitches at manufacturing magnetic varnish for producing hard magnetic disks (HMD). Also we recommend to mix EV and PF according to the ratio 4:1 to be used as binding working layer of HMD because such pair of pitches has least value $k' = 0,14$.

Interaction of molecules of the dissolved substance with each other results in sharp increase of viscosity in the concentrated solutions of polymers comparing with viscosity of the diluted solutions.

Interest to rheological properties of the concentrated solutions of studied mixes EV:PF is caused not only by technological problems of their processing in a product, but also by an opportunity to receive the information on their structure which can be presented as fluctuation grid formed by more or less densely packed units or molecular complexes. Solvent molecules are supposed to be distributed in that grid.

EXPERIMENT

Samples were examined on rotational viscometer «REOTEST» with working unit cylinder – cylinder in an interval of temperatures 298÷338 K. Measurements of viscosity of the concentrated solutions of a mix of epoxynovolaic and phenolformaldehyde pitches have shown that dynamic viscosity [η] decreases with raise in temperature. It testifies that supramolecular structures are destructed under action of thermal motion in studied solutions. Increase of concentration of solutions, on the contrary, increases viscosity and the strain of shift caused by growth of both number and sizes of aggregates. Apparently from fig. 2, the action of second factor prevails at lower temperatures (from 298 up to 313 K), though effect of first factor determines structure of a solution at higher ones (from 318 up to 338 K).

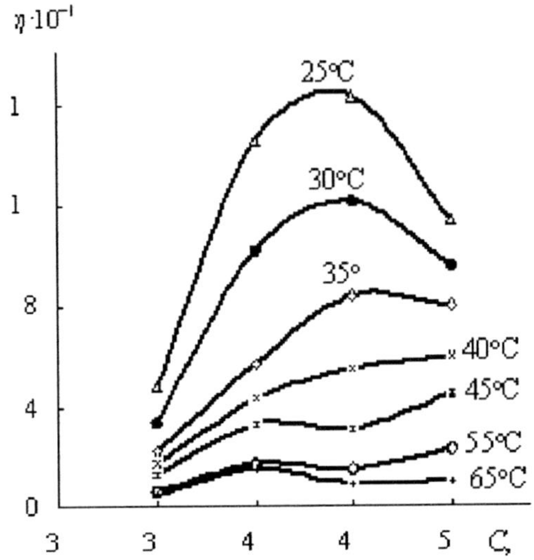

Figure 2. Concentration curve of dynamic viscosity η of solutions of mixed pitches EV:PF = 4:1 in medium isoforon : etiltselozolv = 2:1 at temperatures 25 ÷ 65 °C

Using Frenkel – Eiring formula expressing temperature dependence of viscosity

$$\eta = A\exp\{\Delta G_B / RT\} \tag{3}$$

we have calculated values of activation heat and activation entropy of viscous flux of studied solutions (see Table 1) using kinetic dependence lg $\eta \sim T^{-1}$. ΔG_B in (3) is the free activation energy of viscous flux.

It's seen from these data that there is the durability growth of supramolecular structure with raise in concentration up to 45 g / (100 ml). Such supramolecular structure is formed by molecular aggregates of dissolved polymers. Occurred aggregates also become ordered. The further increase in concentration leads to loosening of structures and to reduction of orderliness.

It follows from results, that for obtaining strong wearproof working layers of hard magnetic disks, in which the mix of epoxynovolaic and phenolformaldehyde pitches is used as binding, it is possible to recommend structures of magnetic varnishes with the content of a polymeric mix not more than 45 g / (100 ml).

Table 1. Activation parameters (ΔH_B and ΔS_B) of viscous flux of mix EV:PF = 4:1

C, g/100 ml	35	40	45	50
ΔH_B, kJ/mol	48,7	63,3	67,8	47,8
ΔS_B, kJ/mol·K	54,8	93,6	108,3	43

The study of laws of formation of mesh structure aiming the choice of an optimal time – temperature mode of solidification of epoxynovolaic pitches, arranged within frameworks of Infra – Red spectroscopy, has shown that there is a formation of a spatial grid in the mechanism of interaction of epoxide groups (EG) of epoxinovolaic pitch with hydroxyl groups of phenolformaldehyde pitch (PFHG) [4].

The structural changes occurring in a polymer matrix during solidification are reflected in spectra of dielectric losses. There is a reduction of height of maxima of dielectric losses with the increase of solidification degree. The decrease of maxima corresponds to the process of vitrification of epoxide polymers and also to their displacement aside heats (fig. 3). Such change of spectra of a dielectric relaxation testifies that with increase in number of units of a chemical spatial grid the distance between crosslinkings decreases, that in turn leads to reduction of number of segments and increases height of a potential barrier of segmental motion, i.e. raises vitrification temperature of a polymer [5,6].

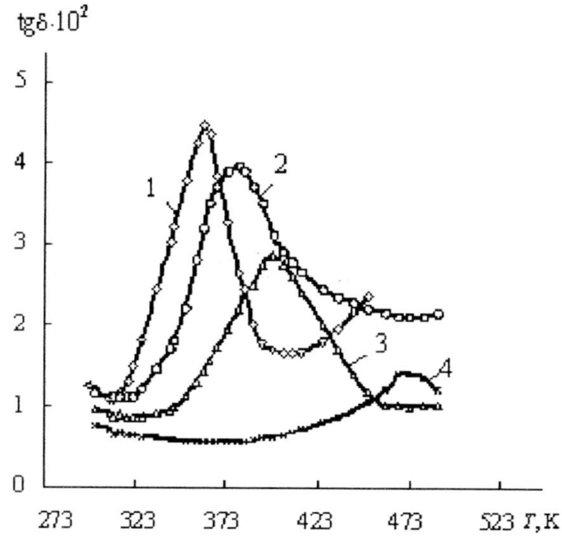

Figure 3. Temperature dependence of tgδ of mix EV:PF = 4:1 of different solidification stages: 1 – stage I; 2 - stage II; 3 – stage III; 4 – stage IV

Discussion

Uniformity of distribution of units of a chemical spatial grid in volume of solidifying mix renders the big influence on thermostability of the investigated polymers.

It's seen from fig. 4 that process of thermooxidative destruction proceeds in three stages for all studied samples excepting one with maximal solidification depth. There were only two steps. Significant (up to 15 %) change of weight at 373-423 K for non – solidified sample on curve of thermogravymetry is associated with evaporation of solvents. Endothermic peak in same temperature interval on curve of differential thermoanalysis confirms this assumption for given sample. Either absence of losses of weight in these temperature areas of solidified samples serves as the certificate to that. The temperature of the beginning of weight losses of studied samples raises from 563 K up to 588 K with growth of a solidification degree, then is reduced, reaching 533 K for a sample with the maximal depth of solidification. In process of growth of a solidification degree I and II stages of decomposition, caused by thermooxidative destruction of aliphatic part of polymer, approach on a temperature scale and merge after the end of solidification process, thus a share of weight losses of stage III grows.

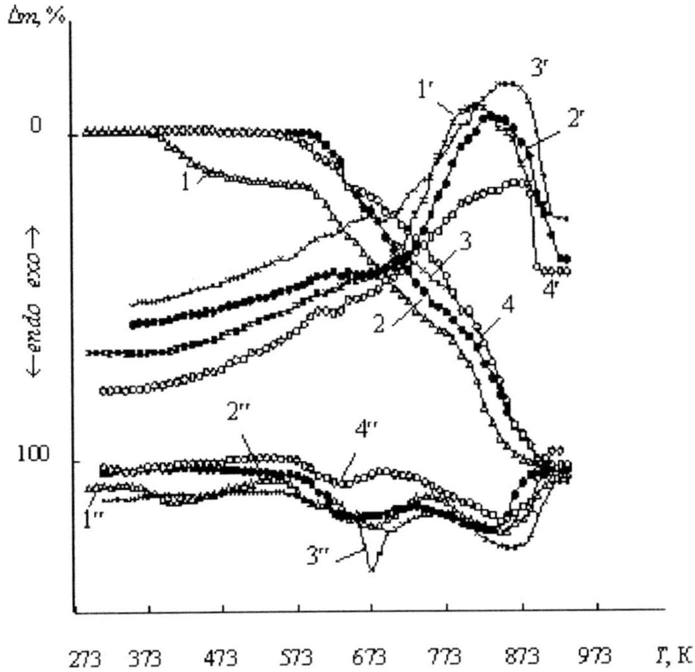

Figure 4. Curves of thermogravymetry analysis (1-4), differential thermoanalysis (1′-4′) and differential thermogravymetry (1″-4″) of mix EV:PF = 4:1 of different solidification stages (I-IV)

The data obtained with help of thermogravymetry testify that there is an increase of density of samples due to growth of linear chains of suprapolymeric structures at first stages of solidification according described mode. Reduction of speed of weight losses of samples also proves that. Thus rise in temperature of the beginning of weight losses of samples is observed. The formation of a spatial grid at a subsequent stages of solidification results in the appearance of heat – setting processes and stressed portions of polymeric chains between chemical units. The increase in an overstrain of such places is apparently the reason of temperature decrease of the beginning of weigh losses of solidified samples. In this connection, it is rational to delay initial stages of solidification while the further process can be proceeded in more severe constraints.

Adsorption – adhesive interactions of polymer with a surface of a filler limiting mobility of its kinetic fragments in a boundary layer results in increase of activation energy of relaxation process in this area and broadening of the spectrum of times of a structural relaxation [7].

Presence of such structural heterogeneity in the polymeric matrix caused by presence of filler, renders considerable influence on processes of structurization during solidification of mixes of epoxynovolaic and phenolformaldehyde pitches [8, 9].

The study of solidification process of epoxynovolaic pitches has shown (at the presence of magnetic particles γ-Fe_2O_3) that filler accelerates the processes of structurization in a researched reactionary mix. And this influence on different stages of solidification is not identical. The influence of filler particles is insignificant on stage II of solidification when processes of growth of molecular chains play the prevailing role. Vice versa presence of filler particles accelerates this process on stage III of solidification when there is intensive formation of a spatial chemical grid. The influence of filler is negligible again on deep degrees of solidification, when the sizes of its particles exceed distance between crosslinkings (IV stage).

Due to upper – said it is reasonable to realize step – like mode of solidification of working layers of the hard magnetic disks containing mix of epoxynovolaic and phenolformaldehyde pitches as polymeric binding. Such a regime avails us to delay process not only on II, but also on III stage of solidification.

Conclusion

On the basis of stated we've offered the solidification mode of a binding working layer of hard magnetic disks on the basis of a mix of epoxynovolaic and phenolformaldehyde pitches:

$$(293\ \text{K}\times 1{,}5\ \text{hour}) \quad \text{(I)}$$
$$+\ (413\ \text{K}\times 3\ \text{hour}) \quad \text{(II)}$$
$$+\ (453\ \text{K}\times 2\ \text{hour}) \quad \text{(III)}$$
$$+\ (473\ \text{K}\times 1{,}5\ \text{hour}) \quad \text{(IV)}$$

It allows to achieve deep solidification degrees and is essential for reducing time of holding a composition at heats (473 K). The performed tests of hard magnetic disks, working layers of which contain a mix of epoxynovolaic and phenolformaldehyde pitches as a binding matrix, solidified on the specified mode have shown that they do not concede to foreign analogues on their physicomechanical and operational characteristics.

References

[1] Ligidov, M.Kh., in Proceedings of Republican Scientific – Practical Conference *"Actual Problems of Chemistry, Biology and Ecology in KBR"*. – (Nalchik, 1997), p. 17-18.

[2] Ligidov, M.Kh., Ulimbasheva, Kh.L. and Guchinov, V.A., in *Abstracts of IV All – Union Conference on Chemistry and Physical – Chemistry of Olygomers*. – (Nalchik, 1990), p. 103.

[3] Ligidov, M.Kh. and Guchinov, V.A., in *Proceedings of Scientific – Practical Conference on Natural Sciences*. – (Nalchik, 1992), p. 21.

[4] Ligidov, M.Kh., Thakahov, R.B. and Mikitaev, A.K., *Plastic Masses*, 12:15 – 16 (2002).

[5] Ligidov, M.Kh., Guchinov, V.A. and Sheriev, A.V., in *Proceedings of Scientific – Practical Conference on Natural Sciences.* – (Nalchik, 1992), p. 26.

[6] Burdin, A.B., Suvorov, A.L., Burdina, L.L., Dulceva, D.L., Honina, T.G. and Sennikov, V.A., *Plastic Masses*, 2:34-36 (2001).

[7] Lipatov, Yu.S., *Fizicheskaja himia napolnennih polimerov*, (Moscow: Chemistry, 1974), 304 p.

[8] Gorbunova, Yu., Kerber, M.L., Balashov, I.N. et al., *Visokomolek. Soed*, 43A:1331-1339 (2001).

[9] Krizhanovsky, V.K., Aladishkin, A.N. et al., *Plastic Masses*, 11:15-18 (2001).

In: Polymers, Polymer Blends, Polymer Composites... ISBN 1-60021-168-2
Eds: A.K. Mikitaev, M.K. Ligidov et al., pp.25-30

Chapter 3

CHAIN FRACTAL GEOMETRY AND DEFORMABILITY OF POLYMER COMPOSITES

Georgi V. Kozlov[1]*, Alexandr I. Burya[2] and Gennadi E. Zaikov[1]
[1]N.M. Emanuel Institute of Biochemical Physics of Russian Academy of Sciences, 119991 Kosygin St. 4, Moscow, Russian Federation
[2]Dnepropetrovsk State Agrarian University, 49600 Voroshilov St. 25, Dnepropetrovsk, Ukraine

ABSTRACT

It is shown, that the fractal geometry of chain part between its fixation points to a great extent is determined by the carbon plastics deformability. This is true also in relation to the density of cluster network of physical entanglements, which is also influenced on the value of fractal dimension of the mentioned chain part. The carbon plastics plasticity is controlled by state of polymeric matrix structure characterized by its fractal dimension.

Keywords: Carbon plastic; polyethylene; short fibres; deformability; plasticity, fractal dimension.

INTRODUCTION

The ability to bear large strains with following full return at stress removal is property displayed at appropriate conditions by actually all substances consisted of long chain macromolecules [1]. What's more, this property is displaed by exclusively materials of such structure. It is important beyond narrow limits of the term "rubber elasticity", by which it is usually designated. The mentioned property acts at swelling of polymeric networks and polymer's deformation, in general not included in the rubber category, for example, semi-

* Correspondence to: Georgi V. Kozlov, N.M. Emanuel Institute of Biochemical Physics of Russian Academy of Sciences, Moscow, Russian Federation, Kosygin st. 4, 119991. mailto: i_dolbin@mail.ru

crystalline polymers, at viscoelastic behaviour of linear polymers in case of flow in liquid or amorphous state. The basis premise of molecular theory of rubber high-elasticity serves assumption, that the stress in rubbers is the result of deformation of covalent network chains, white the contribution of interaction between chains is negligible. Strictly speaking, this is not quite correct even for true rubbers [1] and what's more for polymers in glassy state. The theory of high-elasticity gives the following expression for estimation of limiting draw ratio λ of elastomers [2]:

$$\lambda = n_{st}^{1/2} \tag{1}$$

where n_{st} is statistical segments number on chain part between its fixation points (chemical cross-linking nodes, physical entanglements and so on).

For application of the considered conception to glassy polymers usually a number of empirical assumptions made accounting into consideration much stronger intermolecular in such systems. Edwards and Vilgis [3] offered sliding links conception, which assumes division of chain between entanglements on smaller fragments, which are fixed, but have significant internal freedom. This results to decrease of polymer's limiting strain in comparison with estimated one according to the equation (1). The authors [4] took into consideration physically strictly this factor by introduction in the equation (1) the exponent (D-1) instead of constant parameter ½:

$$\lambda = n_{st}^{D-1} \tag{2}$$

where D is fractal dimension of chain part between its fixation points.

This dimension is varied in limits $1<D\leq 2$ and characterizes internal freedom degree (mobility) of pointed chain part. The purpose of present paper is the application of a model [4] for description of deformability of carbon plastics based on the high density of polyethylene (HDPE).

EXPERIMENTAL

As a polymeric bonding agent HDPE of industrial production (GOST 16-338-85) with crystallinity degree 0.72, estimated by density measurements, and as filler – carbon fibres (CF) of mark Ural-24 were used. The volume content of CF φ_f in carbon plastics is varied in limits 0.038-0.303. A components blending is made on extruder ZSK-30 at the following technological regime: the temperature of the first zone is 408, the second – 413, the third – 428 and the fourth – 423 K. The specimens for studies in the form of dog-bone with base lenght 20 mm were manufactured by the method of compression molding on machine KUASY-100/25. In production process in machine cylinder the following temperature regime was maintained: the first zone – 423, the second – 428 and the third – 443 K. The temperature of casting of specimens was made on testing machine FRZ-100/1 of firm Heckert at the testing temperatures 293, 313, 333 and 353 K and strain rate 10^{-2} s^{-1}.

RESULTS AND DISCUSSION

As the results of mechanical tension testing showed, limiting draw ratio of studied carbon plastics is decreased at the increase of carbon fibres volume content in interval ~ 6.0-1.1. In order to obtain quantitative interrelation of λ and D according to the equation (2), was made calculation D by the following method. The value of fractal (Hausdorff) dimension d_f of carbon plastics structure is determined according to the equation [5]:

$$d_f = (d-1)(1+\nu) \tag{3}$$

where d is dimension of Euclidean space, in which a fractal (obviously, in our case d=3) is considered, ν is Poisson's ratio, the value of which can be estimated on mechanical testing results with the aid of relationship [6]:

$$\frac{\sigma_Y}{E} = \frac{1-2\nu}{6(1+\nu)} \tag{4}$$

where σ_Y is yielding stress, E is elasticity modulus.

Then the value of characteristic ratio C_∞ has been determined, which is an indicator of polymer chain statistical flexibility according to the equation [7]:

$$C_\infty = \frac{2d_f}{d(d-1)(d-d_f)} + \frac{4}{3} \tag{5}$$

Then the relative fraction of local order regions (clusters) φ_{cl} have been estimated with the aid of the equation [7]:

$$d_f = 3 - 6\left(\frac{\varphi_{cl}}{SC_\infty}\right)^{1/2} \tag{6}$$

where S is cross-section area of macromolecule, which is equal to 18.9 Å^2 for polyethylenes.

And at last, the value D can be determined according to the following equation [7]:

$$\frac{2}{\varphi_{cl}^{red}} = C_\infty^D \tag{7}$$

where φ_{cl}^{red} is a reduced value φ_{cl}, taking in to account availability of filler and estimating according to the equation [8]:

$$\varphi_{cl}^{red} = \frac{\varphi_{cl}}{1-\varphi_f} \tag{8}$$

The value n_{st} can be directly calculated from the equation [8]:

$$D = \frac{\ln n_{st}}{\ln(4-d_f) - \ln(3-d_f)} \tag{9}$$

It is obvious, that the parameter n_{st} characterizes density of cluster network of physical entanglements [7]. In Fig. 1 the dependence $\lambda(n_{st})$ is shown, from which the increase λ at n_{st} raising follows, just as was expected from the equations (1) and (2). It is interesting, that this dependence breaks up into two parts with sharply differing slopes, for all that this transition is realized at $n_{st}\approx 9$. As it is known, mentioned value n_{st} is boundary between densely and loosely cross-linked polymers. The data of Fig. 1 suppose, that this classification is true and for cluster network of physical entanglements the value λ changes insignificantly and fracture is quasi-brittle and decrease of network density results to very fast rise λ in narrow interval n_{st}. It is necessary to note, that the form of shown in Fig. 1 dependence $\lambda(n_{st})$ exludes the possibility of application of the equation (1) with constant exponent ½ for its correct description.

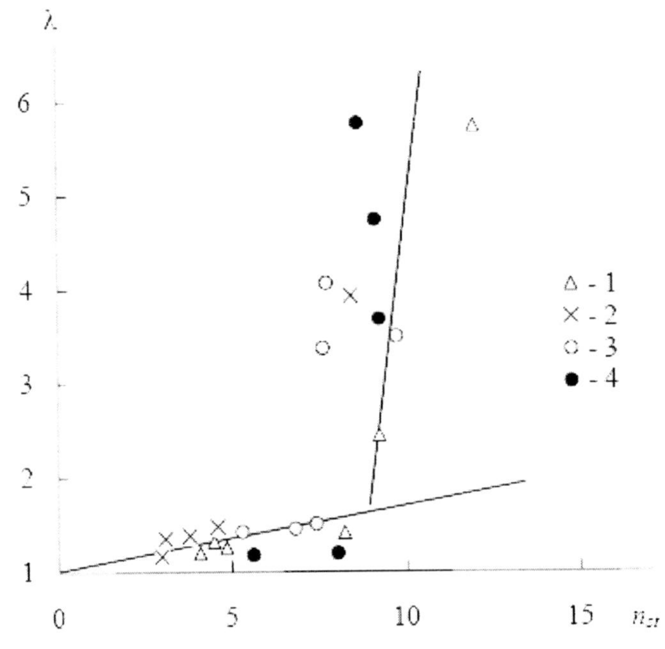

Figure 1. The dependence of limiting draw ratio λ on statistical segments number n_{st} on chain part between clusters for carbon plastics at testing temperatures: 293 (1), 313 (2), 333 (3) and 353 K (4)

In Fig. 2 the comparison experimental λ^e and calculated according to the equation (2) λ^T values of limiting draw ratio for studied carbon plastics at four testing temperatures are given. As can be seen, between λ^e and λ^T enough close correspondence is received and observed scatter of data relationship 1:1 is symmetrical and due to well known statistical character of fracture process.

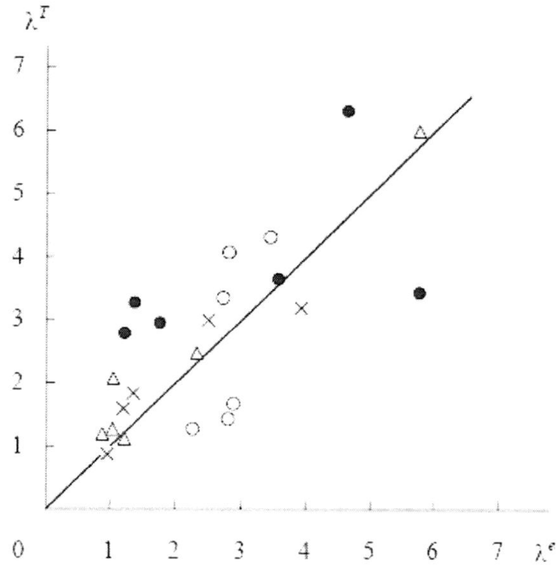

Figure 2. The comparison of experimental λ^e and calculated according to the equation (2) λ^T values of limiting draw ratio for carbon plastics. The straight line give relation 1:1. The notation is than one, that in Fig. 1

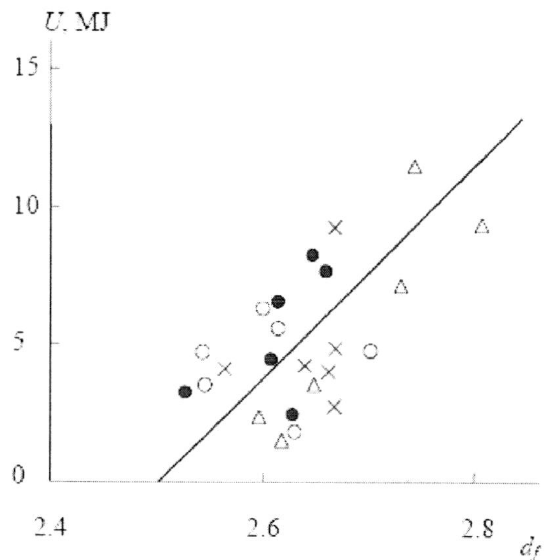

Figure 3. The dependence of fracture work U on structure fractal dimension d_f for carbon plastics. The notation is than one, that in Fig. 1

As it is known, the work of fracture U, characterizing expenditure of energy on material deformation up to failure, is one of the most important plasticity characteristics. In paper [5] it was shown, that the fracture character of solids is determined by the fractal dimension d_f of their structure: at d_f=2.50 brittle fracture is realized, at d_f=2.50-2.67 – quasibrittle (quasiductile) fracture and at d_f>2.70 – ductile fracture. This classification allows to suppose plasticity increase characterized by value U at raising d_f. Actually, the dependence $U(d_f)$ shown in Fig. 3 confirms this assumption. This dependence is linear, the increase U is observed at raising d_f and zero value U is reached at d_f=2.50, i.e., at brittle fracture. Since limiting (maximal) value d_f for real solid is equal to 2.95 [5], then this allows to estimate maximal value U for studied carbon plastics, which is equal to ~ 17 MJ. The scatter of Fig. 3 data is again due to statistical character of fracture process.

Conclusion

Consequently, the results received in present paper showed, that the fractal geometry of chain part between its fixation points to a great extent is determined by the carbon plastics deformability. This is also true in relation to the density of cluster network of physical entanglements, which is also influenced on the value of fractal dimension of the mentioned chain part. The carbon plastics plasticity is controlled by state of polymeric matrix structure characterized by its fractal dimension.

References

[1] Flory, P.J., *Polymer J.*, 17: 1 (1985).

[2] Haward, R.N., Thackray, G., *Proc. Roy. Soc. London*, A 302: 453 (1968).

[3] Edwards, S.F., Vilgis, T., *Polymer*, 28: 375 (1987).

[4] Kozlov, G.V., Serdyuk, V.D., Dolbin, I.V., *Materialovedenie*, 12: 2 (2000).

[5] Balankin, A.S., *Synergetics of Deformable Body*. Moscow, Publisher of Ministry for Defence SSSR, 1991, 404 p.

[6] Kozlov, G.V., Sanditov, D.S., *Anharmonic Effects and Physical-mechanical Properties of polymers*. Novosibirsk, Nauka, 1994, 261 p.

[7] Kozlov, G.V., Zaikov, G.E., *Structure of the Polymer Amorphous State*. Utrecht-Boston, Brill Academic Publisher, 2004, 465 p.

[8] Aloev, V.Z., Kozlov, G.V., *Physics of Orientational Phenomenas in Polymeric Materials*. Nal'chik, Polygraphservis and T, 2002, 288 p.

In: Polymers, Polymer Blends, Polymer Composites... ISBN 1-60021-168-2
Eds: A.K. Mikitaev, M.K. Ligidov et al., pp.31-37

Chapter 4

THE ROLE OF DIFFUSIVE PROCESSES IN MODEL REACTION OF REETHERIFICATION

Lyubov Kh. Naphadzokova* and Georgi V. Kozlov
Kabardino-Balkarian State University, Nal'chik, Russian Federation

ABSTRACT

It is shown, that the offered model of diffusive processes for chemical reactions describes well the main characteristics of model reaction of reetherification. A structure of reaction product (heptylbenzoate molecule) and forming parts of it are the main factor controlling diffusive processes in this case. Mentioned processes are described within the framework of strange (anomalous) diffusion.

Keywords: Chemical reaction, reetherification, scaling, fractional differentiation, active time, strange diffusion.

INTRODUCTION

One of the perspective ways of search of effective inorganic filler-catalysis's for complex polyethers is kinetic study of the reetherification model reaction, performed in the presence of various inorganic compounds [1]. Such method allows to use obtained results in the synthesis process of filled complex polyethers [2].

Synthesis processes in common case can be considered as the complex system self-organization, developing during time, that results to formation of time-dependent fractal structures [3]. In such reactions the important role is played by diffusive processes, which in the considered case have very specific nature. This specificity is due to the fact, that in chemical reactions not all reagents contacts occur with proper for reaction's product

*Correspondence to: Lyubov Kh. Naphadzokova, Kabardino-Balkarian State University, Nal'chik, Russian Federation, Chernyshevski 173, 360004, mailto: lnaph@freemail.ru

formation orientation of reacting molecules. This aspect of reaction is accounted for by steric factor p ($p \leq 1$) [4]. Variation p can result to the change of diffusion type, structure, reaction's product and, as the consequence, to the rate of chemical reaction change. This question can be explained by a simple example. As it is known [5], characteristic size $r(t)$ of region, which can be visited by the reagent molecule during time t, is equal to:

$$r(t) \sim t^{1/(2+\theta)} \tag{1}$$

where θ is connectivity index of reactive medium.

For the case of classical Gaussian diffusion $\theta=0$ and, believing $r(t)=2$ and $t=4$ relative units, the equality within the framework of the relationship (1) will be obtained. Such equality assumes $p=1$, i.e., each contact of reagents molecules results to reaction product formation. Let's assume, that the value p decreases up to 0.05, i.e., only one from 20 contacts of reagents molecules is formed a new chemical sort. This means the increase t in 20 times and then at $r(t)=2$ and $t=80$ relative units from the relationship (1) will be obtained $\theta=4.33$. Since θ is connected with dimension of walk trajectory of reagents molecules d_w by the simple equation [5]:

$$d_w = 2 + \theta \tag{2}$$

then θ increase results to d_w raising, i.e., is slows down the chemical interactions process.

In its turn, the value d_w is connected with Herst exponent H by the equation [5]:

$$d_w = \frac{1}{H} \tag{3}$$

The change θ from 0 up to 4.33 results to raising d_w from 2 (Brownian motion) up to $d_w=6.33$ according to the equation (2) and to reduction H from 0.5 up to 0.158 according to the equation (3). As it is known [5] subdiffusive (slow) transport processes correspond to the values $0 \leq H < 0.5$ and classical Gaussian diffusion – $H=0.5$. Therefore, the decrease p from 1.0 up to 0.05 results to qualitative change of diffusion type too: it changes from Gaussian classical to anomalous (strang). Let's note, that the mentioned transition can occur without changing of general diffusive processes in reactive medium, too since it is due to "rejection" of all diffusive phenomena's, not resulting to the chemical reaction, i.e., to formation of new chemical substance. Proceeding from the said above, the purpose of the present paper is to study diffusive processes influence within the framework of the offered treatment on main characteristics of reetherification model reaction.

EXPERIMENTAL

The reetherification model reaction kinetics of methylbenzoate by heptanole-1 in mica presence was studied at 443 K. Mica catalytic activity was determined on the observed rate constant of first order k_1 at the twentieth multiple of heptanole-1 excess and mica contents 30 mass. % in calculation on the methylbenzoate [2].

The reetherification kinetics was studied on the gas chromatograph "Biokhrom" with using as internal standard diphenyloxide according to the earlier described method [1]. The rate constant k_1 was calculated according to the equation of irreversible reaction of the first order.

The mica flagopit with polydispersity 0.749 and average probable particles size 0.23×10^{-6} m is used. The initial mica (conditional designation NMM) and also mica chemically modified by sodium hydroxide (SMM) and sulphur acid (AMM) were applied.

RESULTS AND DISCUSSION

Earlier it was shown [6], that for reaction of type

$$A + B \rightarrow inert \quad product \tag{4}$$

the scaling relationship is true:

$$\rho_A \sim t^{D/4} \tag{5}$$

where ρ_A is concentration "surviving" in the reaction process particles, t is reaction duration, D is dimension controlling the reaction elapsion.

In case of reaction elapsion in the Euclidean spaces the value D is equal to dimension of this space d and for fractal spaces D is accepted equal to spectral dimension d_s [6]. By graphing ρ_A=(1-Q) (where Q is conversion degree) as a function of t in double logarithmic coordinates the value D from the slope of these graphs can be determined. It was found, that the mentioned graphs fall apart on two linear parts: at t<100 min with small slope and at t>100 min the slope essentially increases. In this case the value d_s varies within the limits 0.069-3.06. Since the considered reactions are elapsed in Euclidean space, that is pointed by a linearity of kinetic curves Q-t, this means, that the reetherification reaction elapses in specific medium with Euclidean dimension d, but with connectivity degree, characterized by spectral dimension d_s, typical for fractal spaces [5].

The authors [5] have formulated fractional equation of transport processes, having the following form:

$$\frac{\partial^\alpha \psi}{\partial t^\alpha} = \frac{\partial^{2\beta}}{\partial r^{2\beta}}(B\psi) \tag{6}$$

where $\psi=\psi(t, r)$ is distribution function of particles, $\partial^{2\beta}/\partial r^{2\beta}$ is Laplacian operator in d – dimensional Euclidean space and B is relation of transport generalized coefficient and d. The introduction of fractional derivatives $\partial^\alpha/\partial t^\alpha$ and $\partial^{2\beta}/\partial r^{2\beta}$ allows to account for the effects of memory (α) and nonlocality (β) in context of common mathematical formalism [5].

The introduction of fractional derivative $\partial^{\alpha}/\partial t^{\alpha}$ in the kinetic equation (6) allows for random walks in fractal time (RWFT) – "time component" of strange dynamical processes in turbulent mediums [5]. The distinctive feature of RWFT serves the absence of any noticeable jumps in particles behaviour; in this case root-mean-square displacement $\langle r^2(t)\rangle$ increases with t as t^{α}. The parameter α has sense of fractal dimension of "active" time, in which real walks of particles look as random process; interval of active time is proportional to t^{α} [5].

In its turn, the exponent 2β in the equation (6) allows for instantaneous jumps of particles (Levy "flights") from one region to another. Therefore, exponent's relation α/β gives relation of RWFT contact frequencies and Levy "flights". The value β in the first approximation can be adopted as constant and then relation α/β will be inversed proportional to waiting time of chemical reaction realization. The value α/β is equal to [5]:

$$\frac{\alpha}{\beta}=\frac{d_s}{d} \tag{7}$$

In Fig. 1 the dependence Q on α/β for boundary times of the mentioned above parts of dependences ln(1-Q)-ln t (t=60 and 300 min) is shown. As can be seen, linear correlation $Q(\alpha/\beta)$, passing through coordinates origin and assuming raising Q at increase α, is obtained. Therefore, the larger active time t^{α} the more intensively reaction elapses, that, in general has been exputed.

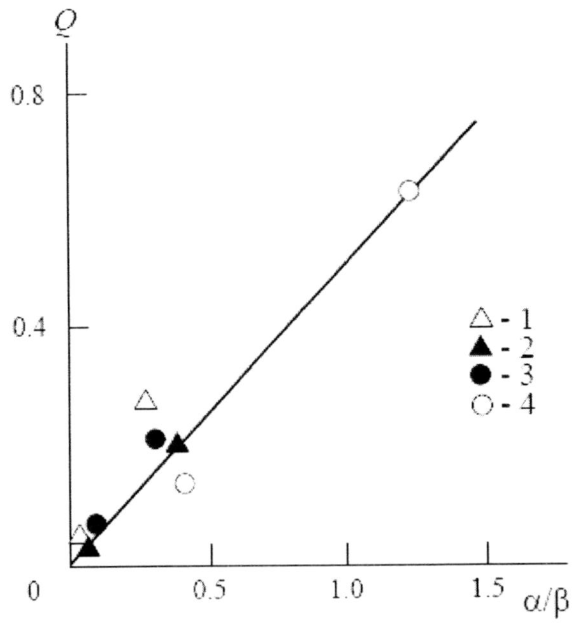

Figure 1. The dependence of conversion degree Q at t=60 and 300 min on relation of the equation (6) exponents α/β for reetherification reaction without mica (1) and in presence of NMM (2), SMM (3), AMM (4)

Next we'll be consider the influence on diffusive processes in the course of chemical reaction of the steric factor p, the value of which can be estimated according to the equation (at t=10600 s) [7]:

$$p = \frac{1.6}{10600^{(D_f - 1)/2}} \tag{8}$$

where D_f is dimension of reetherification product (molecule of heptylbenzoate), determined with the aid of the equation [8]:

$$t^{(D_f - 1)/2} = \frac{C_1}{k_1(1 - Q)} \tag{9}$$

where C_1 is constant, determined according to the boundary conditions and adopted in present paper equal to 8×10^{-4} s^{-1}.

In Fig. 2 the dependence $\mu^{1/2}=(\alpha/\beta)^{1/2}$ on p (such form of correlation was chosen with the purpose of its linearization) is shown. As can be seen, correlation $(\alpha/\beta)^{1/2}(p)$ is actually linear and passes through the origin of coordinates. These points, that, as has been assumed above, diffusive processes in the course of reetherification reaction are controlled by the probability of new chemical substance formation, i.e., molecule of heptylbenzoate.

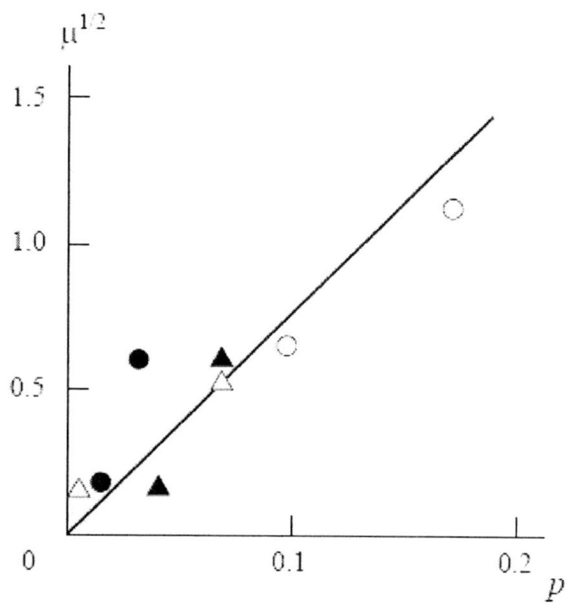

Figure 2. The dependence of the exponent μ on steric factor p for reetherification reaction. The notation is the same, as in a Fig. 1

Since μ=2H [5], then according to the equation (3) can be calculated the value d_w or, more strictly, effective value d_w. In Fig. 3 the dependence $d_w(p)$ is shown, which has very nonlinear form, conditionally dividing into three parts. For small p<0.025 large values $d_w \approx$10-60 are obtained and, correspondingly, large values $D_f \approx$2.13. This value D_f corresponds to chemically-limited mechanism of cluster-cluster aggregation (D_f=2.11), which is characterized by small probability of clusters sticking together or, in other words, by small values p [9]. At p=0.025-0.10 the interval d_w=2-10 and value D_f reduces in average up to 1.71, that corresponds to diffusive-limited mechanism of cluster-cluster aggregation (D_f=1.75) [9]. And at last, the decrease d_w smaller than 2, i.e., the approach of particle trajectories to linear with d_w=1, results to formation of heptylbenzoate molecule with D_f<1.5. This value D_f corresponds to transparent macromolecular coil dimension [10], i.e., at such dimension molecules of reagents can freely pass through each other, that noticeably facilitates the elapsion of reetherification reaction.

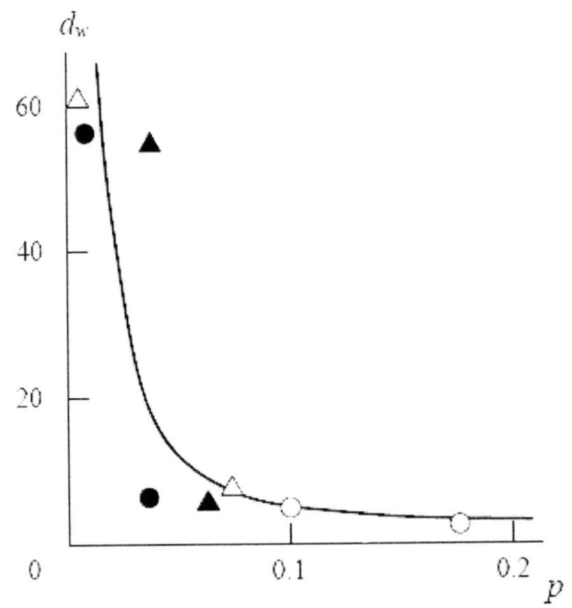

Figure 3. The dependence of walk trajectory of reagents particles d_w on steric factor p for reetherification reaction. The notation is the same, as in a Fig. 1

CONCLUSION

Thus, the results of the present work showed, that the offered model of diffusive processes for chemical reactions describes well the main characteristics of model reaction of reetherification. A structure of reaction product (heptylbenzoate molecule) and forming parts of it is the main factor controlling diffusive processes in this case. Mentioned processes are described within the framework of strange (anomalous) diffusion.

REFERENCES

[1] Naphadzokova, L.Kh., Vasnev, V.A., Tarasov, A.I., *Plast. massy*, 3: 39 (2001).

[2] Vasnev, V.A., Naphadzokova, L.Kh., Tarasov, A.I., Vinogradova, S.V., Lependina, O.L., *Vysokomolek. Soed. A*, 42: 2065 (2000).

[3] Karmanov, A.P., Matveev, D.V., Monakov, Yu.B., *Doklady AN*, 380: 635 (2001).

[4] Barns, F.S., *Biophysica*, 41: 790 (1996).

[5] Zelenyi, L.M., Milovanov, A.V., *Uspekhi Fisichesk. Nauk*, 174: 809 (2004).

[6] Meakin, P., Stanley, H.E., *J. Phys. A*, 17: L173 (1984).

[7] Kozlov, G.V., Shustov, G.B., *in "Uspehki v Oblasti Fiziko-Khimii Polimerov"*. Moscow, Khimia, 2004, 341 p.

[8] Kozlov, G.V., Bejev, A.A., Lipatov, Yu.S., *J. Appl. Polymer Sci.*, 92: 2558 (2004).

[9] Smirnov, B.M., *Uspekhi Fisichesk. Nauk*, 149: 177 (1986).

[10] Baranov, V.G., Frenkel', S.Ya., Brestkin, Yu.V., *Doklady AN SSSR*, 290: 369 (1986).

In: Polymers, Polymer Blends, Polymer Composites… ISBN 1-60021-168-2
Eds: A.K. Mikitaev, M.K. Ligidov et al., pp.39-58

Chapter 5

THERMAL DEGRADATION AND COMBUSTION OF POLYPROPYLENE NANOCOMPOSITE

S. M. Lomakin[1*], I. L. Dubnikova[2], S. M. Berezina[2], G. E. Zaikov[1], R. Kozlowski[3], Gyeong-Man Kim[4] and G. H. Michler[4]

[1]N.M. Emanuel Institute of Biochemical Physics of Russian Academy of Sciences, 119991 Kosygin 4, Moscow, Russia

[2]N.N. Semenov Institute of Chemical Physics of Russian 119991 Kosygin 4, Moscow, Russia

[3]Institute of Natural Fibres, Poznan, ul. Wojska Polskiego 71 b, Poland

[4]Martin-Luther-Universität Halle-Wittenberg, Geusaer Straße, D-06217 Merseburg, Germany

ABSTRACT

Polypropylene (PP) has wide acceptance for use in many application areas. However, low thermal resistance complicates its general practice. The new approach in thermal stabilization of PP is based on the synthesis of PP nanocomposites. This paper discusses new advances in the study of the thermo-oxidative degradation of PP nanocomposite. The observed results are interpreted by a proposed kinetic model, and the predominant role of the one-dimensional diffusion type reaction. According to the kinetic analysis, PP nanocomposites had superior thermal and fireproof behavior compared with neat PP. Evidently, the mechanism of nanocomposite flame retardancy is based on shielding role of high-performance carbonaceous-silicate char which insulates the underlying polymeric material and slows down the mass loss rate of decomposition products.

Keywords: charring; diffusion; flammability; kinetics; nanocomposite; polypropylene; thermal degradation

[*]Correspondence to: Sergei M. Lomakin, N.M. Emanuel Institute of Biochemical Physics ofRussian Academy of Sciences, Moscow, Russia, Kosygin 4, 119991, mailto:lomakin@sky.chph.ras.ru

INTRODUCTION

Polypropylene (PP) has wide acceptance for use in many application areas. However, low thermal and flame resistance complicates its general practice.

Much interest has been devoted to the study of PP thermal degradation [1-5]. Earlier studies on the thermal breakdown of polypropylene and its mechanism have been reviewed by Bockhorn et al [1]. The thermal depolymerization of polypropylene was considered as a radical process, which includes initiation, propagation and termination, which is analogous to the mechanism of polyethylene degradation [1]. After the bond scission into primary and secondary radicals, tertiary radicals are formed *via* rearrangements reactions. Subsequent β-scission leads to volatile alkenes and the chain carriers [1]. A β-scission to the other side leads to a short secondary radical and a polymer chain with a terminated double bond. This short secondary radical is saturated *via* intramolecular hydrogen transfer and results in an alkane. Due to the low alkane concentration, transfer reactions seem to play a minor role in polypropylene degradation. The apparent kinetic parameters for the overall thermal degradation of PP were determined under isothermal conditions using a gradient free reactor with on-line mass spectrometry [1]. For pure PP, different investigators have reported values of activation energy from 220–270 kJ/mol [1-3].

Chan and Balke applied thermogravimetric data for the pyrolysis of PP to provide a kinetic model of thermal degradation [4]. However, the wide temperature range used in this work (45–580°C) encompassed a change in the decomposition mechanism and this greatly limited the utility of the methods. To accommodate this mechanism change, the data were treated as a pseudo first-order reaction [4]. It was observed that the data conformed to a first-order fit at temperatures of less than 404 to 421°C (depending upon heating rate) with activation energy of 98.3 ± 3.1 kJ/mol [4]. At higher temperatures the data could again be fit as a pseudo first-order reaction, but with an activation energy of 327.9 ± 8.6 kJ/mol. The two regions were separated by a relatively narrow transition region. The lower activation energy occurring at lower degradation temperatures is attributed to scission of 'weak links' in the polymer. The higher activation energy was similar to the carbon-carbon bond dissociation energy and is associated with random scission throughout the polymer [4].

In contrast with many previous publications, Gao et al. showed that a first-order reaction model cannot be applied to describe thermal degradation of polypropylene [5]. The appropriate reaction order was determined to be 0.35 by using the degree of conversion at the maximum degradation rate measured under dynamic conditions. The validity of the reaction order thus determined was verified by similarity of activation energy between the single heating rate plot and the isoconversional plot. The reaction order was also supported by the consistency of the Arrhenius parameters of isothermal degradation with dynamic degradation [5].

The new approach in thermal stabilization of PP is based on polymer nanotechnology [6, 7]. Over the past decade, polymer nanocomposites have received considerable interest as an effective way for developing new composite materials, and they have been studied widely. Because of the larger surface area and surface energy of the additives when individual particles become smaller, it is not an easy task to obtain homogeneously dispersed organic / inorganic composites when the additives are down sized to nano-scale [8]. Melt intercalation has been successful in preparing polymer clay nanocomposites [9].

Recently the thermal degradation behavior of nanocomposites based upon PP-organoclay was studied by Zanetti, Camino *et. al.* using isothermal and dynamic thermogravimetry [10]. They suggested that the oxygen charring action and scavenging effect in the nanocomposite increases as the volatilization proceeds and that in the nanocomposite a catalytic role is played by the intimate polymer-silicate contact that may further favor the oxidative dehydrogenation-crosslinking-charring process.

This silicate morphology may act as an efficient barrier to oxygen diffusion towards the bulk of the polymer. Surface polymer molecules trapped within the silicate are thus brought to a close contact with oxygen to produce the thermally and oxidative stable charred material providing a new char-layered silicate nanocomposite acting as an effective surface shield [10]. In the present study a routine set of TGA analytical data was accomplished to provide the kinetic analysis of thermal degradation in air of PP nanocomposite based on layered organoclay (Cloisite 20A). The results obtained in this study gave an additional evidence of diffusion-controlled character of thermal degradation of PP nanocomposite caused by the catalytic-charring effect of nanosilicate clay.

Lately Qin at al. reported data on polypropylene/montmorillonite (PP/MMT) microcomposites thermal degradation and flammability [11]. They mentioned that PP microcomposites exhibit higher thermal stability and considerably reduced peak heat release rate due to physico-chemical adsorption of the volatile degradation products on the silicates [11]. On the other hand, the addition of MMT can catalyze the initial decomposition of PP matrix and accelerate its ignition in combustion. It has been observed that a ceramic-like char formed on the surface of the composites during the combustion test. The characterization of the char surface prior to ignition indicates that it is an inorganic-rich surface, leading to improvement of the thermal stability and reduction of flammability of the composites because of its better barrier properties [11].

Experimental

Synthesis

Polypropylene (PP) by Moscow petroleum refinery (MFI = 0.7) and maleic anhydride-modified oligomer (MAPP - Licomont AR 504 by Clariant co.) with M_n~2900; MA-content~4 wt.%. were blended/mixed using a laboratory Brabender mixing chamber for 2 min. at the first stage. The concentration of MAPP was 20% by weight of original PP. Then the 7% wt. of organoclay (Cloisite 20A - Na^+ montmorillonite modified by dimethyl, dihydrogenatedtallow ammonium chloride by Southern Clay Co.) was added to the PP-MAPP - melt at a rotor speed of 60 rpm and set temperature of 190°C (PP-MAPP- Cloisite 20A). 10 min. mixing time was used in all the experiments.

Characterization

WAXS analysis of nanocomposite layered structure was carried out with a DRON-2 X-ray Diffractometer with Cu-Kα radiation. Diffraction patterns were collected in reflection-mode geometry from 2° to 10°2θ.

AFM studies were performed with commercial scanning probe microscope Nanoscope IIIA and IV MultiMode (Digital Instruments/Veeco Metrology Group, USA) in tapping mode, at ambient conditions. Conventional etched Si probes (stiffness ~40 N/m, resonant frequency 160-170 kHz) were used. The amplitude of the free-oscillating probe, A_0, was varied in the 10-20 nm range while the set-point amplitude, A_{sp}, ranged from 0.5 to 0.8 A_0. Imaging was conducted on the flat sample surfaces being prepared at –80°C with an ultramicrotome MS-01 (MicroStar Inc., USA) equipped with a diamond knife.

The thermal decomposition studies were performed over a temperature range of 20-600°C using a MOM Q 1500 thermogravimetric analysis (TGA) system under air environment at the scan rates of 3, 5 and 10K/min.

Kinetic analysis of PP compositions thermal degradation was carried out using Thermokinetics software by NETZSCH-Gerätebau GmbH.

Results and Discussion

Structure Characterization

It is always necessary to carefully characterize the polymer structure in order to ensure a sort of the dispersion for the nanoclays in polymers. XRD analysis and TEM would provide some information on the nanocomposite structural morphology. The diffraction patterns for nanoclay and nanocomposites are displayed in Figure 1 (a). The Cloisite 20A itself has a single peak at around 3.6° with d-space of 2.4 nm.

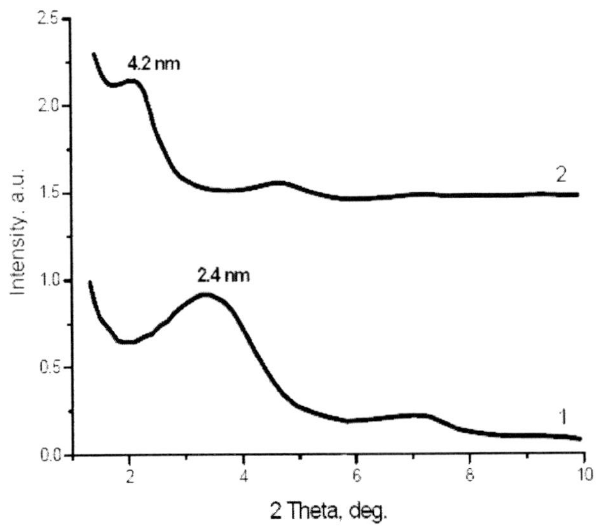

Figure 1. WAXS analysis for Cloisite 20A (1) and PP-MAPP-Cloisite 20A (2)

The shift of the clay basal spacing d_{001} from 3.6° to 2.2° in PP-MAPP-Cloisite 20A sample suggests the intercalated nanocomposite sample have higher d-space (4.2 nm) than that in the original clay (2.4 nm), it may have some exfoliated structures considering the smearing of peak in nanocomposite sample.

However, the XRD can only detect the periodically stacked montmorillonite layers (A); for all these nanocomposites there also exists a small number of exfoliated layers (B) as well, which can be directly observed by transmission electron microscopy (TEM).

In Figure 2 we present TEM images with different magnification which designate the presence of intercalated tactoids (A) and apparently exfoliated monolayers (B) coexisting in the nanocomposite structure. The intercalated structures are characterized by a parallel registry that gives rise to the XRD reflection of Fig. 1.

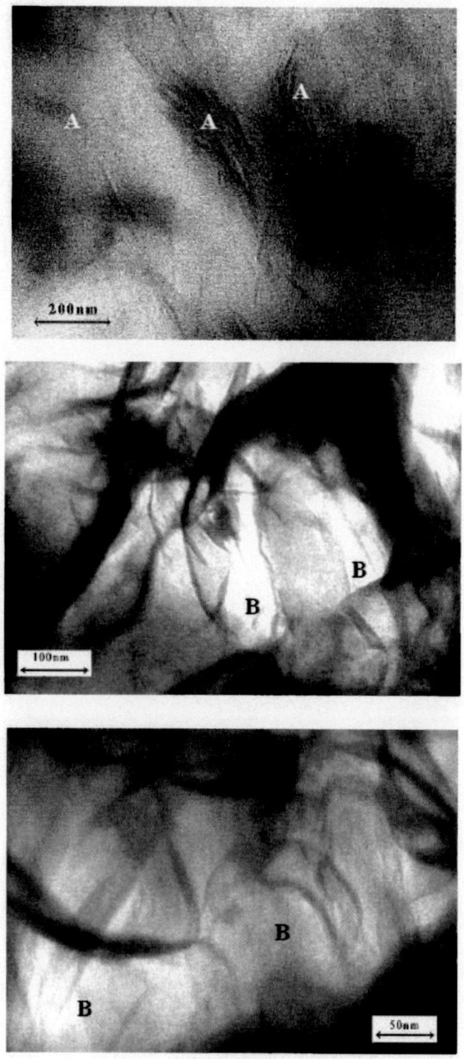

Figure 2.TEM photos with different magnification of PP-MAPP-Cloisite 20A where: A - stacks of layers (intercalated tactoids), B - exfoliated monolayers

AFM studies demonstrate the similar morphology characteristics for PP-MAPP-Cloisite 20A (Fig. 3). Height and phase images were simultaneously recorded on polymer surfaces. Height image presents surface topography, whereas phase images provide a sharp contrast of fine structural features and emphasize differences in sample components. We presume the availability of essentially intercalated structure with the complex multilayered tactoids (A) and small amount of exfoliated monolayer units (B) (Fig. 3).

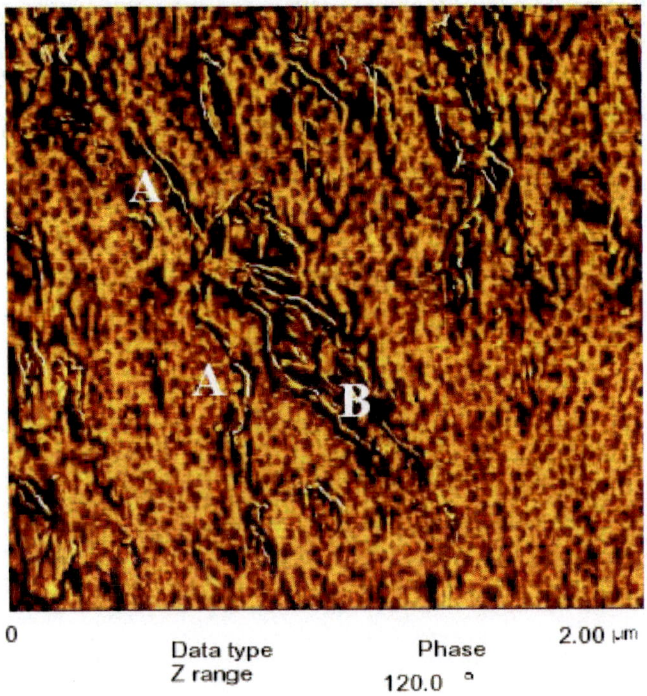

Figure 3. AFM image of PP-mPP with 7% Cloisite where: A - stacks of layers (intercalated tactoids), B-exfoliated monolayers

Thermo-Oxidative Degradation Study

TG analysis of PP and PP nanocomposite (PP-MAPP-Cloisite 20A) show that at heating in air at 10°C/min, PP volatilizes completely, in two steps beginning at about 300°C with maximum rate at 400°C through a radical chain process propagated by carbon centered radicals originated by carbon-carbon bond scission (Fig.4) [12]. Below 200°C, the hydroperoxidation on C-H bonds, in which oxygen addition occurs to the carbon radicals created within the polymer chain by H abstraction initiates radical-chain degradation of PP [13], whereas above 200-250°C, oxidative dehydrogenation of PP takes over [14]. Depolymerization and random scission by direct thermal clevege of carbon-carbon bonds becomes possible in air as in nitrogen, above 300°C.

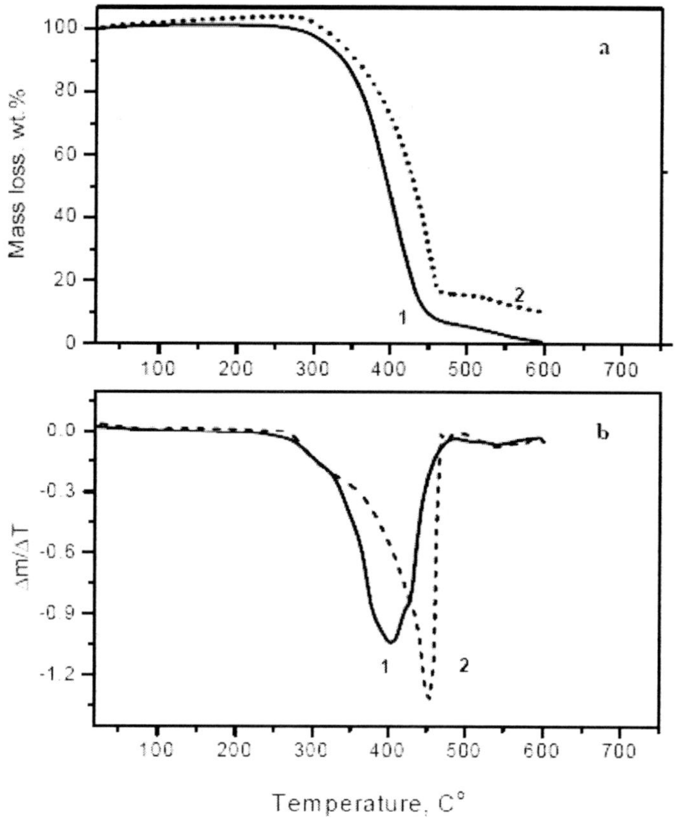

Figure 4. TG – (a) and DTG – (b) curves of PP (1) and PP-MAPP-Cloisite 20A (2) in air at the heating rate of 10K/min

Stabilizing effect of Δ50°C (Fig.4 b) of PP-MAPP-Cloisite 20A over neat PP calculated with the maximum rate of mass loss can be explain by means of the barrier effect of the silicate nanolayers which operate in the nanocomposite level against oxygen diffusion, shielding the polymer from its action.

A char residue from neat PP is left at 450°C (5%), due to charring promoted by oxidative dehydrogenation. Then it slowly decomposes on heating up to 600°C in air (Fig.4). On the other hand, thermal degradation of PP-MAPP-Cloisite 20A in air results to much more stable char form which doesn't oxidize even at 600°C (Fig. 4 and 5).

Silicate nanostructure executes a role of efficient barrier to oxygen diffusion towards the native polymer. Surface polymer molecules trapped within the silicate are thus brought to a close contact with oxygen and catalytic - silicate layers to produce the thermally and oxidative steady carbonized structures (Scheme on Fig.5).

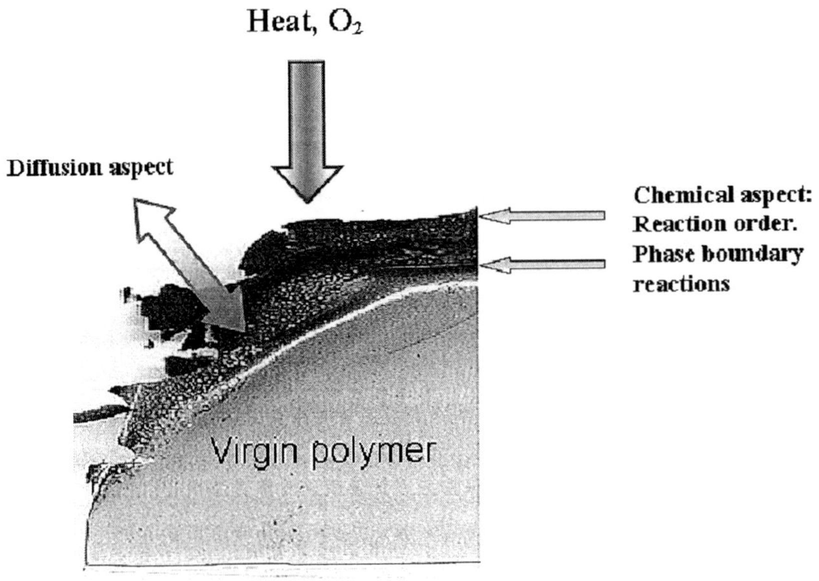

Figure 5. Scheme of the thermo-oxidative degradation of PP nanocomposite (PP-MAPP-Cloisite 20A)

Kinetic Analysis Using TGA Data

Kinetic studies of materials degradation have been carried out for many years using numerous techniques to analyze the data. Most often, TGA is the experimental method of choice and the only technique to be explored here. TGA involves placing a sample of polymer on a microbalance within a furnace and monitoring the weight of the sample during some temperature program. It is generally accepted that materials degradation obeys the basic equation (1) [15]

$$dc/dt = - F(t,T\ c_o\ c_f) \tag{1}$$

where: t - time, T - temperature, c_o - initial concentration of the reactant, and c_f - concentration of the final product. Equation $F(t,T,c_o,c_f)$ can be described by two separable functions, $k(T)$ and $f(c_o,c_f)$:

$$F(t,T,c_o,c_f) = k(T(t)\cdot f(c_o,c_f) \tag{2}$$

Arrhenius equation (4) will be assumed to be valid for the following:

$$k(T) = A\cdot\exp(-E/RT) \tag{3}$$

Therefore,

$$dc/dt = - A\cdot\exp(-E/RT)\cdot f(c_o,c_f) \tag{4}$$

A series of reactions types: classic homogeneous reactions and typical solid state reactions, is listed in Table 1 [15].

Table 1. Reaction types and corresponding reaction equations,
$dc/dt = -A \cdot exp(-E/RT) \cdot f(c_o,c_f)$

Name	$f(c_o, c_f)$	Reaction type
F_1	c	first-order reaction
F_2	c^2	second-order reaction
F_n	c^n	n[th]-order reaction
R_2	$2 \cdot c^{1/2}$	two-dimensional phase boundary reaction
R_3	$3 \cdot c^{2/3}$	three-dimensional phase boundary reaction
D_1	$0.5/(1 - c)$	one-dimensional diffusion
D_2	$-1/ln(c)$	two-dimensional diffusion
D_3	$1.5 \cdot e^{1/3}(c^{-1/3} - 1)$	three-dimensional diffusion (Jander's type)
D_4	$1.5/(c^{-1/3} - 1)$	three-dimensional diffusion (Ginstling-Brounstein type)
B_1	$c_o \cdot c_f$	simple Prout-Tompkins equation
B_{na}	$c_o^n \cdot c_f^a$	expanded Prout-Tompkins equation (na)
$C_{1\text{-}X}$	$c \cdot (1+K_{cat} \cdot X)$	first-order reaction with autocatalysis through the reactants, X. $X = c_f$.
$C_{n\text{-}X}$	$c^n \cdot (1+K_{cat} \cdot X)$	n[th]-order reaction with autocatalysis through the reactants, X
A_2	$2 \cdot c \cdot (-ln(c))^{1/2}$	two-dimensional nucleation
A_3	$3 \cdot c \cdot (-ln(c))^{2/3}$	three-dimensional nucleation
A_n	$N \cdot c \cdot (-ln(c))^{(n-1)/n}$	n-dimensional nucleation/nucleus growth according to Avrami/Erofeev

The analytical output must fit of measurements with different temperature profiles by means of a common kinetic model.

Kinetic analysis of PP compositions thermo-oxidative degradation at heating rates of 3, 5 and 10K/min. was carried out using NETZSCH Thermokinetics software in order to provide an extra evidence of the diffusion-stabilizing effect of nanoclay structure.

Model-free methods evaluations were chosen as the starting points in kinetic analysis of neat PP and PP-mPP with 7% Cloisite 20A for determining the activation energy in the development of the model. Figure 6 shows a corresponding Friedman analysis, where the activation energy is a function of partial mass loss change [16].

The curves show higher values at the beginning of the sintering process, i.e. at lower partial-length-change values, and considerably higher values at the end of the process (a) and particularly (b). This indicates the presence of a multiple-step process.

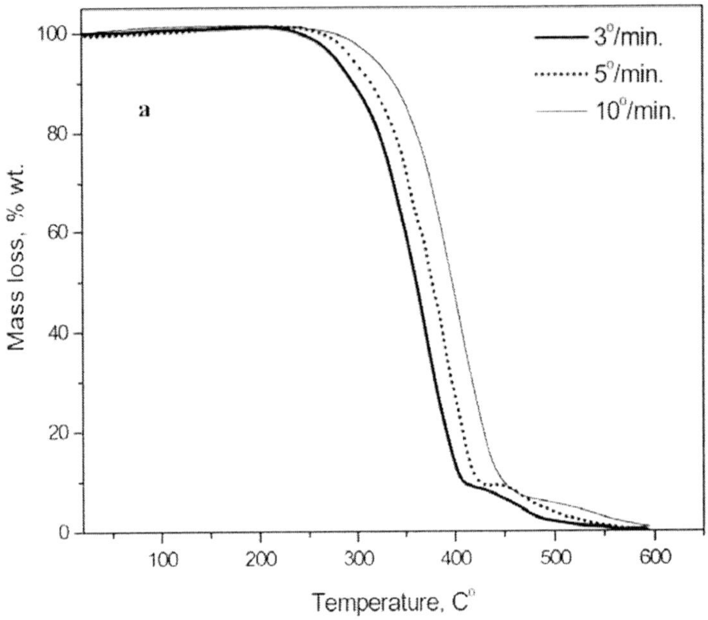

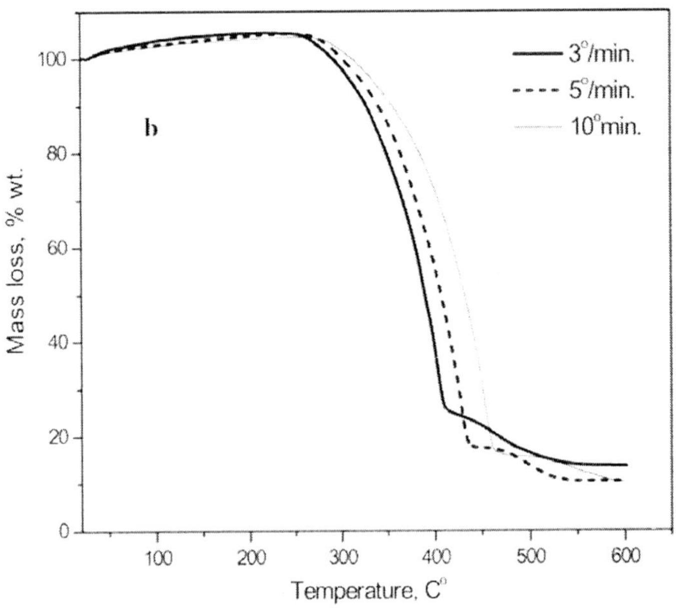

Figure 6. Friedman Analysis of neat PP – (a) and PP-MAPP-Cloisite 20A – (b)

First round analysis by Friedman method indicates a complexity of the scheme for neat PP and PP-MAPP-Cloisite 20A thermal degradation in air [16]. Nonlinear fitting procedure established the two - stage scheme for neat PP

$$A \rightarrow X_1 \rightarrow B \rightarrow X_2 \rightarrow C \quad (4)$$

and the triple - stage scheme for PP-mPP with 7% Cloisite 20A (Fig. 4 a, b) [15,17].

$$A \rightarrow X_1 \rightarrow B \rightarrow X_2 \rightarrow C \rightarrow X_3 \rightarrow D \quad (5)$$

Taking these findings into consideration for neat PP, a fit was attempted using nonlinear regression with model (4), where the nth-order (F_n) reaction type was used for all steps of the reaction (Fig. 7, Table 2).

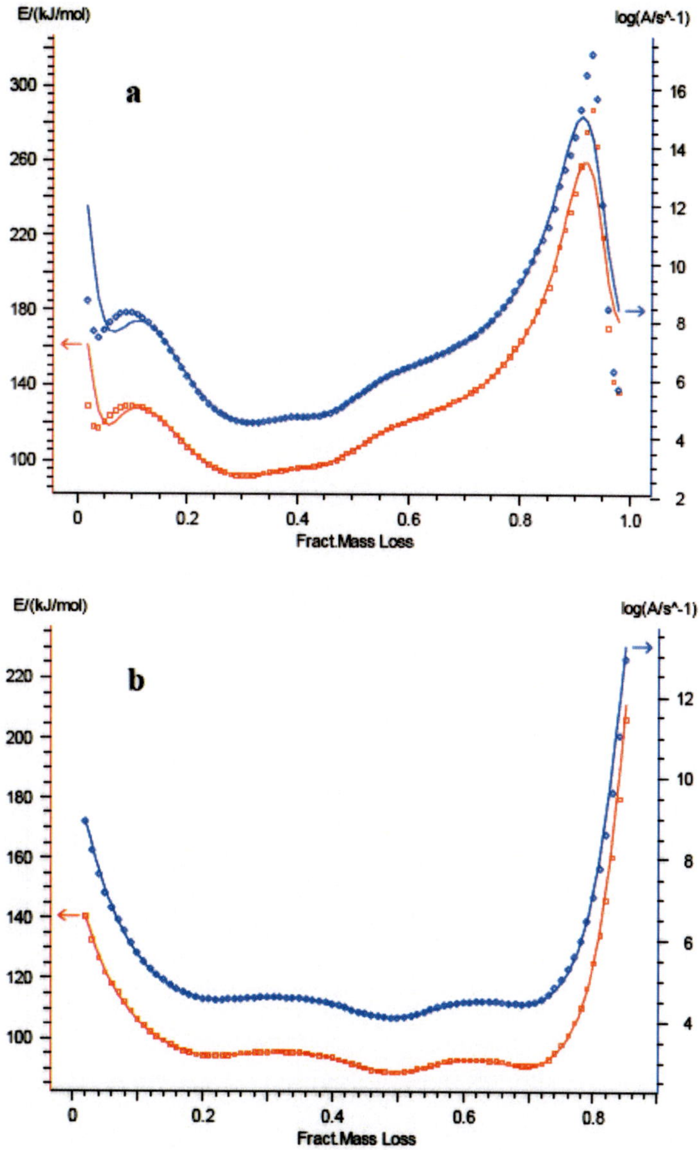

Figure 7. Nonlinear kinetic modelling for neat PP. Comparison between experimental TG data (dots) and model results (lines) at the heating rates of 3, 5 and 10K/min

With this expanded model, an excellent fit is possible for all three measurements. The kinetic parameters are listed in Table 2.

Table 2. Kinetic parameters resulting from multiple-curve analyses (heating rates 3, 5 and 10 K/min) with reaction model ($A \rightarrow X_1 \rightarrow B \rightarrow X_2 \rightarrow C$) from TG measurement of neat PP

Reaction models (types, X*i*)	Parameter	Value	Corr. Coeff.
$F_n \rightarrow F_n$	$\log A_1$, s^{-1}	6.4	0.9994
	E_1, kJ/mol	110.3	
	n_1	1.13	
	$\log A_2$, s^{-1}	9.9	
	E_2, kJ/mol	151.6	
	n_2	2.59	

A more sophisticated model (5), based on different reaction types, was chosen for PP-MAPP-Cloisite 20A thermo-oxidative degradation. The parameters are listed in Table 3.

Table 3. Kinetic parameters resulting from multiple-curve analyses (heating rates 3, 5 and 10 K/min) with different reaction models ($A \rightarrow X_1 \rightarrow B \rightarrow X_2 \rightarrow C \rightarrow X_3 \rightarrow D$) from TG measurement of PP-MAPP-Cloisite 20A

Reaction models (types, X*i*)	Parameter	Value	Corr. Coeff.
$F_n \rightarrow F_n \rightarrow F_n$	$\log A_1$, s^{-1}	6.3	0.9974
	E_1, kJ/mol	113.4	
	n_1	1.16	
	$\log A_2$, s^{-1}	8.8	
	E_2, kJ/mol	150.9	
	n_2	2.46	
	$\log A_3$, s^{-1}	11.5	
	E_3, kJ/mol	188.5	
	n_3	0.78	
$F_n \rightarrow D_1 \rightarrow F_n$	$\log A_1$, s^{-1}	6.9	0.9988
	E_1, kJ/mol	113.4	
	n_1	1.21	
	$\log A_2$, s^{-1}	4.7	
	E_2, kJ/mol	100.0	
	$\log A_3$, s^{-1}	12.0	
	E_3, kJ/mol	199.8	
	n_3	1.17	

Table 3. Continued

<table>
<tr><th>Reaction models (types, Xi)</th><th>Parameter</th><th>Value</th><th>Corr. Coeff.</th></tr>
<tr><td rowspan="3">$F_n \rightarrow D_2 \rightarrow F_n$</td><td>$logA_1$, s^{-1}
E_1, kJ/mol
n_1</td><td>6.4
113.3
1.68</td><td rowspan="3">0.9974</td></tr>
<tr><td>$logA_2$, s^{-1}
E_2, kJ/mol</td><td>6.2
118.4</td></tr>
<tr><td>$logA_3$, s^{-1}
E_3, kJ/mol
n_3</td><td>11.7
197.2
0.95</td></tr>
<tr><td rowspan="3">$F_n \rightarrow D_3 \rightarrow F_n$</td><td>$logA_1$, s^{-1}
E_1, kJ/mol
n_1</td><td>6.5
113.6
2.04</td><td rowspan="3">0.9973</td></tr>
<tr><td>$logA_2$, s^{-1}
E_2, kJ/mol</td><td>8.3
152.3</td></tr>
<tr><td>$logA_3$, s^{-1}
E_3, kJ/mol
n_3</td><td>11.8
197.1
0.94</td></tr>
<tr><td rowspan="3">$F_n \rightarrow D_4 \rightarrow F_n$</td><td>$logA_1$, s^{-1}
E_1, kJ/mol
n_1</td><td>6.6
113.7
2.01</td><td rowspan="3">0.9973</td></tr>
<tr><td>$logA_2$, s^{-1}
E_2, kJ/mol</td><td>7.2
138.9</td></tr>
<tr><td>$logA_3$, s^{-1}
E_3, kJ/mol
n_3</td><td>11.9
195.2
0.98</td></tr>
<tr><td rowspan="3">$F_n \rightarrow A_2 \rightarrow F_n$</td><td>$logA_1$, s^{-1}
E_1, kJ/mol
n_1</td><td>6.6
114.4
2.10</td><td rowspan="3">0.9975</td></tr>
<tr><td>$logA_2$, s^{-1}
E_2, kJ/mol</td><td>5.6
105.2</td></tr>
<tr><td>$logA_3$, s^{-1}
E_3, kJ/mol
n_3</td><td>12.0
199.3
1.23</td></tr>
<tr><td rowspan="3">$F_n \rightarrow A_3 \rightarrow F_n$</td><td>$logA_1$, s^{-1}
E_1, kJ/mol
n_1</td><td>6.5
114.1
2.02</td><td rowspan="3">0.9976</td></tr>
<tr><td>$logA_2$, s^{-1}
E_2, kJ/mol</td><td>4.8
95.3</td></tr>
<tr><td>$logA_3$, s^{-1}
E_3, kJ/mol
n_3</td><td>12.4
200.1
1.46</td></tr>
</table>

Table 3. Continued

Reaction models (types, X*i*)	Parameter	Value	Corr. Coeff.
$F_n \rightarrow A_n \rightarrow F_n$	$\log A_1$, s^{-1}	6.7	0.9983
	E_1, kJ/mol	114.7	
	n_1	0.97	
	$\log A_2$, s^{-1}	4.7	
	E_2, kJ/mol	97.7	
	$\log A_3$, s^{-1}	11.9	
	E_3, kJ/mol	200.0	
	n_3	1.26	
$F_n \rightarrow R_2 \rightarrow F_n$	$\log A_1$, s^{-1}	7.1	0.9979
	E_1, kJ/mol	114.9	
	n_1	1.27	
	$\log A_2$, s^{-1}	5.1	
	E_2, kJ/mol	105.7	
	$\log A_3$, s^{-1}	12.1	
	E_3, kJ/mol	200.3	
	n_3	1.76	
$F_n \rightarrow R_3 \rightarrow F_n$	$\log A_1$, s^{-1}	7.2	0.9978
	E_1, kJ/mol	114.6	
	n_1	1.10	
	$\log A_2$, s^{-1}	5.2	
	E_2, kJ/mol	108.9	
	$\log A_3$, s^{-1}	11.6	
	E_3, kJ/mol	188.5	
	n_3	0.78	

Taking these fittings for PP-MAPP-Cloisite 20A, a best approximation was attempted using nonlinear regression with model (5), based on the best fit quality (correlation coefficient) (Fig. 8), where the one-dimensional diffusion (D_1) reaction type was used for the second step of the reaction (Table 3), whereas the nth-order reaction models were chosen for the first and third steps respectively.

These results show that the second step in thermo-oxidative degradation of PP-MAPP-Cloisite 20A is described by one-dimensional diffusion (D_1) reaction type which is liable for the overall process of the carbonization in nanocomposite polypropylene structure.

Flame Resistant Properties

The recent interest in the reported char-promoting functionalized dispersed nanoclays to yield nanocomposite structures having enhanced fire and mechanical properties, when the clays are present only at levels of 2~10%, prompts their investigation as potential fire retardants. Because of its wholly aliphatic hydrocarbon structure, neat polypropylene by itself burns very rapidly with a relatively smoke-free flame and without leaving a char residue. It has a high self-ignition temperature (570°C), a rapid decomposition rates and hence has a high flammability.

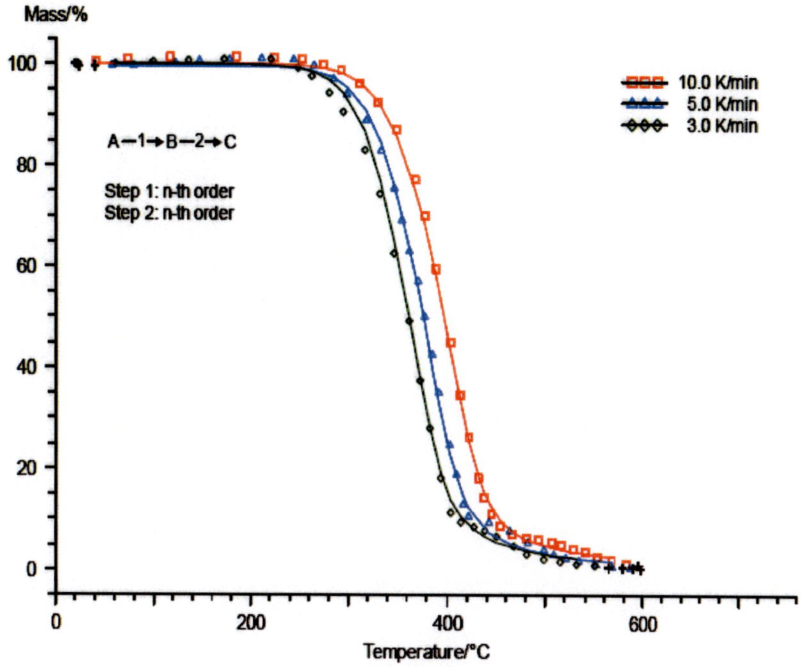

Figure 8. Nonlinear kinetic modelling for PP-MAPP-Cloisite 20A. Comparison between experimental TG data (dots) and model results (lines) at the heating rates of 3, 5 and 10K/min

Polypropylene nanocomposites have attracted more and more interest in flame retardant area in recent years due to their improved fire properties [18-20]. It is suggested that the presence of clay can enhance the char formation providing a transient protective barrier and hence slowing down the degradation of the matrix [19, 20].

The kinetic results of the present study let us the starting point to predict the mass loss of material under isothermal pyrolysis conditions using the same thermokinetics software. Figure 9 shows the fractional mass loss curves as a function of time with temperature (400 – 600°C) as a parameter.

It is clearly seen that under conditions of polymer ignition and initial surface combustion, the mass loss for PP-MAPP-Cloisite 20A and its rate are noticeably lower then adequate values for the neat PP. An improvement in flame resistance of PP-MAPP-Cloisite 20A over the neat PP happens as a result of the char formation providing a transient protective barrier. In the present study this phenomena was interpreted in terms of isothermal kinetic analysis.

Apart from this information, the graphs of mass loss rates (*dm/dt*) vs. time for neat PP and PP-MAPP-Cloisite 20A indicate the depression of the degradation (fuel) products under the isothermal pyrolysis conditions at 600°C (Fig. 10). It is well known that the temperature of 600°C corresponds to an incident heat flux of 35 kW/m^2; this is referred to the real scale fire scenario [21].

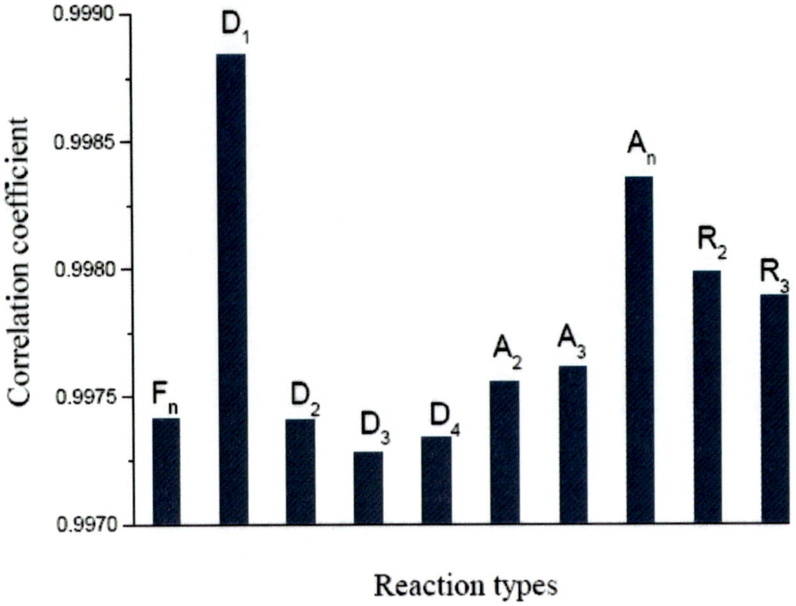

Figure 9. Fractional reaction vs. time for neat PP (a) and PP-MAPP-Cloisite 20A (b)

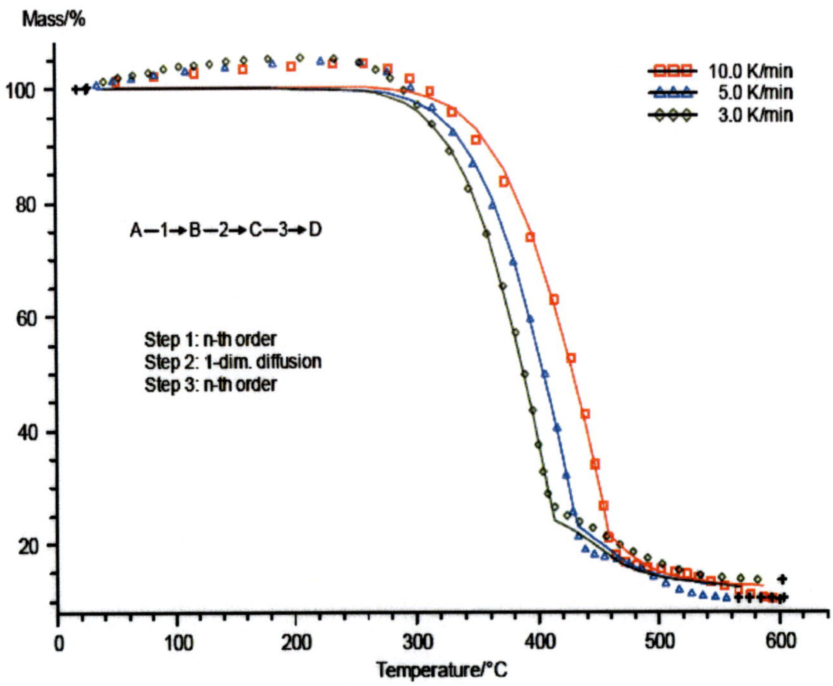

Figure 10. Mass loss rates vs. time for neat PP-MAPP-Cloisite 20A (2) under the isothermal heating condition of 600°C

The cone calorimeter is one of the most effective bench-scale methods for studying the flammability properties of materials. Fire-relevant properties, measured by the cone calorimeter, such as heat release rate (RHR), maximum RHR, smoke and carbon monoxide yield, are vital to the evaluation of the fire safety of materials [22].

In the present study the combustibility of polypropylene nanocomposite was evaluated by a cone calorimeter. The tests were performed at an incident heat flux of 35 kW/m^2 using the cone heater [21]. Peak heat release rate (RHR), mass loss rate (MLR), specific extinction area (SEA) data, carbon monoxide and heat of combustion data measured at 35 kW/m2, are presented in Figs. 11 and 12.

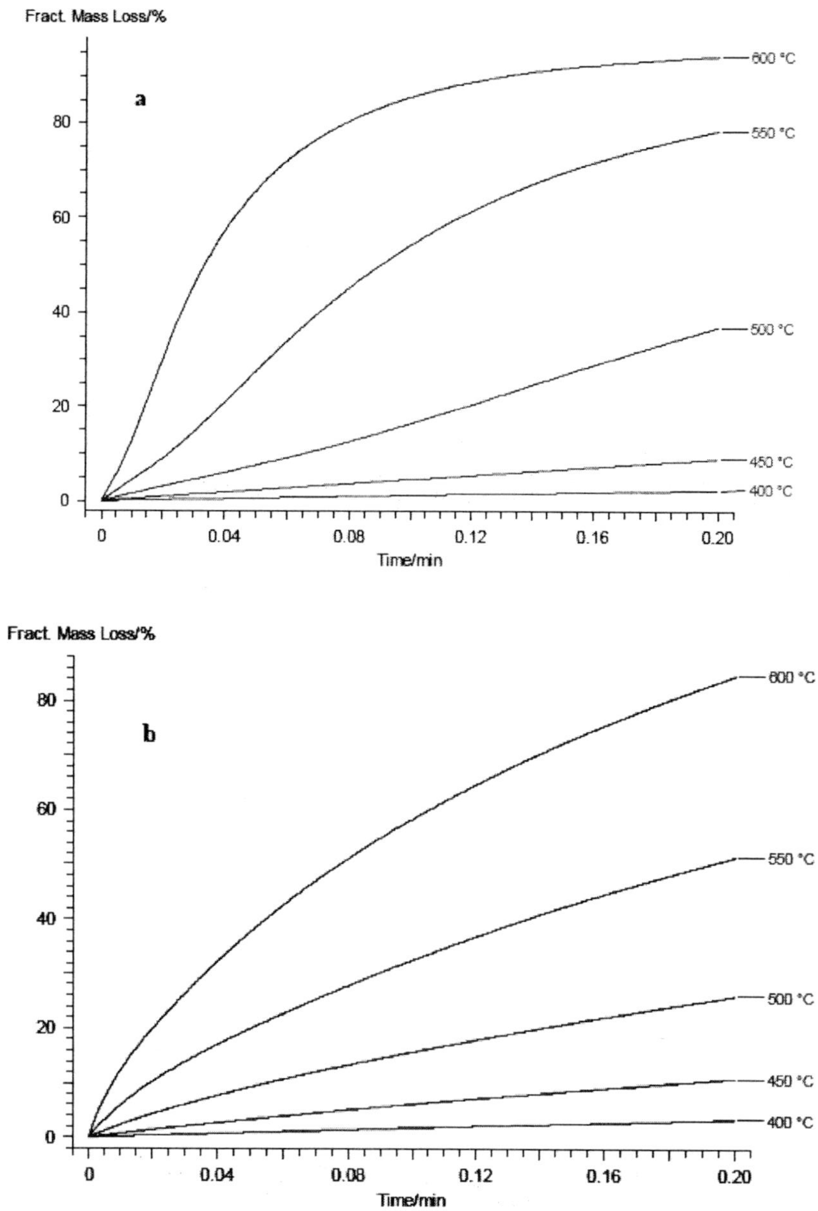

Figure 11. Rate of heat release vs. time for PP and PP-MAPP-Cloisite 20A

The RHR plots for PP-MAPP-Cloisite 20A nanocomposite and PP at 35 kW/m^2 heat flux shown in Figure indicate a 60% - decrease of peak of RHR (Fig.11). Comparison of the Cone calorimeter data PP and PP-MAPP- 7% Cloisite 20A reveals that the specific heat of combustion (H_c), specific extinction area (SEA), a measure of smoke yield, and carbon monoxide yields are practically unchanged; this suggests that the source of the improved flammability properties of these materials is due to differences in condensed-phase decomposition processes and not to a gas-phase effect. The primary parameter responsible for the lower RHR of the nanocomposites is the mass loss rate (MLR) during combustion, which is significantly reduced from the value observed for the pure PP (Fig.12). It is supposed, that this effect is caused by ability to initiate the formation of char barrier on a surface of burning polymeric nanocomposites that drastically limits the heat and mass transfer in a burning zone.

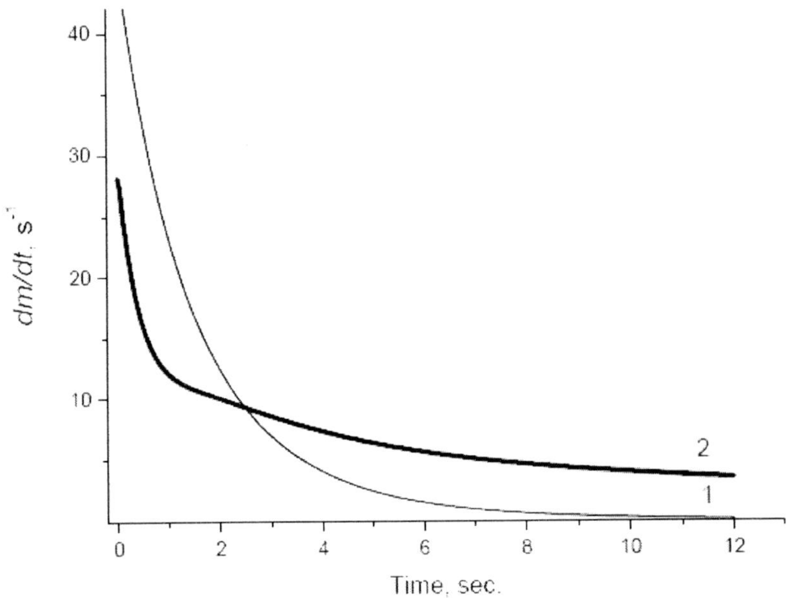

Figure 12. Mass loss rate vs. time for PP and PP-MAPP-Cloisite 20A

Conclusion

The kinetic data obtained by dynamic TGA designate thermal stabilization effect of nanoclay structure into a polymer matrix, caused in one-dimensional diffusion process of catalytic-charring throughout the thermal degradation of PP nanocomposition. According to provided kinetic analysis, polypropylene nanocomposite demonstrated the transcendent thermal and fireproof behaviour in relation to neat polypropylene.

Based on modern concepts of the mechanisms of polymer nanocomposites flame retardancy and char formation, it is possible to assume that simulation of dynamic behavior of anisotropic particles with the different types of morphology and relationship of geometric dimensions presented in viscous polymeric melt will be one of the promising trends in this field of research.

On the other hand, the unique barrier properties of nano-dispersed polymeric composites are of interest of polymer combustion due to specific laminar morphology. This type of structure is especially effective in comparison with the other forms of fillers because of the "labyrinth effect". Researches in this area will allow defining an influence of diffusion of low-molecular products of pyrolysis on the process of micro-intumescence in a superficial layer of burning polymers.

Acknowledgements

This work was supported by the Russian Foundation for Basic Research, project N 04-03-32052.

The authors are pleased to acknowledge the Moscow division of NETZSCH-Gerätebau GmbH for Thermokinetics software.

References

[1] Bockhorn, H, A. Hornung, A, Hornung, U, Schawaller, D, *Journal of Analytical and Applied.*

[2] Pyrolysis, 48(2):93 (1999).2. Ballice, L, Reimert, R, *Chemical Engineering and Processing,* 41:289 (2002).

[3] Murty, MVS, Rangarajan, P, Grulke, EA, Bhattacharyya, D, *Fuel Processing and Technology,* 49:75 (1996).

[4] Chan, JH, Balke, ST, *Polymer Degradation and Stability,* Volume 57:135 (1997).

[5] Gao, Z, Kaneko, T, Amasak,i I, Nakada, M, *Polymer Degradation and Stability,* 80:269 (2003).

[6] Giannelis, E, *Adv Mater*, 8:29 (1996).

[7] Gilman, JW., Kashiwagi, T, Nyden, MR, Brown, JET, Jackson, C L, Lomakin, SM, Giannelis, E P, Manias, E, in *Chemistry and Technology of Polymer Additives*, Chapter 14, ed by Ak-Malaika S, Golovoy A, Wilkie CA, p 249, Blackwell Science Inc., Malden MA (1999).

[8] Zanetti, M, Lomakin, S, Camino, G, *Macromol Mater Eng*, 279:1-9 (2000).

[9] Kojima, Y., Usuki, A., Kawasumi, M., Okada, A., Fukushima, Y., Kurauch,i T., Kamigaito O., *J Mater Res*, 8:1185 (1993).

[10] Zanetti, M, Camino, G, Reichert, P, Mülhaupt, R, *Macromolecular Rapid Communications* 22:176-180 (2001).

[11] Qin, H, Zhang, S, Zhao, C, Feng, M, Yang, M, Shu, Z, Yang, S, *Polymer Degradation and Stability*, 85:807 (2004).

[12] Grassie, N, Scott, G, in *Polymer degradation and stabilization*, Cambridge University Press, Cambridge, p 275 (1985).

[13] March, J, in *Advanced Organic Chemistry*, McGraw-Hill Kogakusa Ltd, Tokyo, p.367 (1977).

[14] Benson, SW, Nogia, PS, *Account of Chemical Research*, 12:233 (1979).

[15] Opfermann, J, *J Thermal Anal Cal*, 60: 641 (2000).

[16] Friedman, HL, *J Polym. Sci.*, C6:175 (1965).
[17] Opfermann, J, Kaisersberger, E, *Thermochim Acta*, 11:167 (1992).
[18] Lomakin, SM, Zaikov, GE, in *Modern Polymer Flame Retardancy*, VSP Int. Sci. Publ. Utrecht, Boston, p. 272 (2003).
[19] Gilman, JW, *Applied Clay Sci*, 15:31 (1999).
[20] Gilman, GW, Jackson, C L., Morgan, A B, Harris, R H, Manias, E, Giannelis, E P, Wuthenow, M, Hilton, D, Phillips, S, *Chem. Mater*, 12:1866 (2000).
[21] Babrauskas, V, *Fire and Materials*, 19:243 (1995).
[22] Babrauskas, V., Peacock, R. D. *Fire Safety J.,* 18:255 (1992).

In: Polymers, Polymer Blends, Polymer Composites... ISBN 1-60021-168-2
Eds: A.K. Mikitaev, M.K. Ligidov et al., pp.59-67

Chapter 6

Fundamental Aspects of Filling of Nanocomposites with High-Elasticity Matrix: Fractal Models

***Georgi V. Kozlov*[1], *Yurii G. Yanovskii*[1]* *and Gennadi E. Zaikov*[2]**

[1]Institute of Applied Mechanics of Russian Academy of Sciences, 119991, Leninskii pr. 32 A, Moscow, Russian Federation

[2]N.M. Emanuel Institute of Biochemical Physics of Russian Academy of Sciences, 119991 Kosygin St. 4, Moscow, Russian Federation

Abstract

In the present paper are considered three fundamental factors controlling filling processes of nanocomposites with high-elasticity matrix. They are: filler particles aggregation, strain localization and change of structure type from Euclidean object up to fractal. The last postulate is the most physically fundamental, since it demonstrates noncorrectness of description of structure and properties of nanocomposites with high-elasticity matrix within the framework of classical theory of entropic high-elasticity.

Keywords: Nanocomposite; high-elasticity matrix; filler; aggregation; strain localization; fractal structure.

Introduction

The practical importance and complexity of structure of polymer composites with matrix, which is in high-elasticity state, are due to a large number of works, devoted to studies of structure and properties of the mentioned composites. Lipatov [1] summarized obtained

*Correspondence to: Yurii G. Yanovskii, Institute of Applied Mechanics of Russian Academy of Sciences, Moscow, Russian Federation, Leninskii pr. 32 A, 119991. mailto: i_dolbin@mail.ru

results and came to the conclusion that the main factors of reinforcement of the considered composites are:

(1) the features of polymer chemical constitution and strong intermolecular interaction of it with filler particles surface;
(2) the formation of second structures by filler particles;
(3) reorganization at deformation of filled material molecular structure without its failure;
(4) the failure of network structure – rupture of filler – polymer bonds or molecular network transverse bonds.

Among other factors, influenced on the reinforcement, Lipatov marks form and size of filler particles, character of their distribution and aggregation and also a number of common physical-chemical causes, typical for all polymer composites [1].

DISCUSSION

Not rejecting correctness of these principles, becoming already classical, we nevertheless distinguished three fundamental factors, the role of which happens obvious owing to the fractal analysis application for the description of polymer composites structure and properties [2-4].

The first of these factors is filler particles aggregation. The detailed description of this effect influence on structure and properties of composites, having glassy matrix, is given in review [3]. For the considered composites this effect is important first of all for determination of filler volume fraction φ_f, which is calculated by division of filler mass fraction into its density. However, filler particles in real composites exist in the form of fractal aggregates, which density can be more smaller than monolithic object of such filler. Below the example of value φ_f calculation for concrete composite will be given.

The second aspect of filler particles aggregation is the possibility of taking into consideration occlusive rubber in aggregates of these particles. As it is known [5], the relationship between fractal aggregate radius R_{ag}, and particles number N in it will look like:

$$R_{ag} \sim N^{1/d_f} \tag{1}$$

where d_f is fractal dimension of filler particles aggregate.

The rise of occlusive rubber fraction in aggregate at condition R_{ag}=const means decrease N and reduction d_f. This gives the possibility of quantitative calculation of occlusive rubber fraction in aggregate.

And at last, the third important aspect of aggregation process is its mechanism. As it is known [6], within the framework of irreversible aggregation models are possible two aggregation mechanisms: cluster-cluster and particle-cluster, the physical sense of which follows from its name. Within the framework of the first mechanism the value R_{ag} is determined by the relationship [3]:

$$R_{ag} \sim \left(\frac{4c_0kT}{3\eta m_0}\right)^{1/d_f} t^{1/d_f} \tag{2}$$

where c_0 is initial concentration of filler particles, k is Boltzmann constant, T is temperature, η is medium viscosity, m_0 is initial particle mass, t is aggregation duration.

For particle-cluster mechanism the value R_{ag} is determined like this [6]:

$$R_{ag} \sim c_0^{-1/(d-d_f)} \tag{3}$$

where d is dimension of Euclidean space, in which a fractal (obviously, in our case d=3) is considered.

From the relationship (2) follows, that at cluster-cluster aggregation (merging of small aggregates in largerones) the value R_{ag} can be controlled by variation of parameters T, η, m_0, d_f and t. But the most interesting is c_0 role, i.e., filler content, in regulation of value R_{ag}: for cluster-cluster mechanism the c_0 rise results to increase R_{ag} and for particle-cluster mechanism – to its reduction. In this aspect the most interesting is the relationship (3), since from it follows, that at definite conditions the aggregation process can be suppressed by c_0. Therefore, the ability of aggregation mechanism change is the perspective instrument of control of structure of filler particles aggregates. Besides, the value d_f ($\approx$ 2.50) for particle-cluster aggregation mechanism is essentially more than corresponding dimension ($d_f \approx$1.80-2.10 [6]) for cluster-cluster mechanism, i.e., the first of the mentioned mechanisms allows to obtain more compact aggregates, practically not containing occlusive rubber.

The second important factor is the calculation of draw ratio λ for composites. This problem is due to heterogeneity of composite structure, consisting of two phases elasticity modulus of one of them is essentially (in some orders of value) more, than the second. Differently speaking, during composite tension process only one phase is extended, namely, rubber, owing to nominal value λ not already does describe adequately composite deformation process. This effect is well known and for its quantitative accounting for there are two models. The first of them uses the equation [7]:

$$\lambda_{mol} = \frac{\lambda}{1-\varphi_f} \tag{4}$$

where λ_{mol} is molecular draw ratio.

The equation (4) recommended itself well at the description of extrudates of componors based on the ultrahighmolecular polyethylene [2]. The second model uses more precise equation for calculation of local draw ratio of two-phase polymeric material [8]:

$$\lambda = \lambda_p(1-\varphi_f) + \lambda_f\varphi_f \tag{5}$$

where λ_p and λ_f are draw ratio for polymeric matrix and filler, accordingly.

Returning to the question of filler particles aggregation, it should be noted, that for compact aggregates of high-modulus filler λ_f=1 and then from comparison of the equations (4) and (5) follows, that for large enough λ $\lambda_{mol}\approx\lambda_p$. However, this situation can be essentially changed for loose (having rather small dimension d_f) aggregates, containing of occlusiv rubber significant fraction. For them λ_f>1 can be expected, that reduces the value λ_p at λ=const, i.e., weakens reinforcing the action of the filler.

And at last, the third and the most fundamental factor is the change of nanocomposite structure at the introduction of particulate filler in high-elasticity polymeric matrix. As Balankin showed [9], classical theory of entropic high-elasticity has a number of principal deficiencies due to non-fulfilment for real rubbers of two main postulates of this theory, namely, essentially non-Gaussian statistics of real polymeric networks and lack of co-ordination of postulates about Gaussian statistics and incompressibility of elastic materials. Last postulate means, that Poisson's ratio ν of these materials must be equal to 0.5. As it is known [10], Gaussian statistics of macromolecular coil is correct only in case of its dimension D_f=2.0, i.e., for coil in θ-solvent. Since between value D_f and fractal dimension d_f^p of polymer's condensed state structure exists the relationship [11]:

$$d_f^p = 1.5 D_f \tag{6}$$

then for execution of chains Gaussian statistics condition in volume polymer the criterion d_f^p=3 realization is required, i.e., the polymer structure must be Euclidean object. If the last condition with definite provisos is carried out for nonfilled rubbers, then the introduction in rubber filler with significantly smaller Poisson's ratio ν_f will decrease the value ν of composite as system. Since the value of fractal dimension of composite structure d_f^c is determined like that [12]:

$$d_f^c = (d-1)(1+\nu) \tag{7}$$

then this means, that the classical entropic high-elasticity theory is applied only to Euclidean objects (ν=0.5, d_f^c=3.0) and, accordingly, inapplicable for description of rubbers, filled by particulate filler, of which ν<0.5 and d_f<d=3, i.e., which are fractal objects. Therefore, for description of such objects behaviour the fractal theory of elasticity and entropic high-elasticity should be used [9].

In order to demonstrate the correctness of expounded above general postulates, let's make quantitative estimations of the dependence of stress σ on φ_f and T for rubber SKI-3 filled by technical carbon P-234 according to the data of paper [13]. For this let's make approximated, by plausible assumption, that the technical carbon particles with diameter d_{part}=50 mnm [13] form the aggregates from five particles. Then the radius of such aggregate can be estimated according to the equation [14]:

$$R_{ag} = \left(\frac{n_{part} S}{\pi \eta_p} \right)^{1/2} \tag{8}$$

where n_{part} is particles number in aggregate, S is cross-sectional area of particle, η_p is packing density, which is equal to 0.74 for monodisperse circles [15].

At n_{part}=5 and d_{part}=50 mnm R_{ag}≈65 mnm. The density of such aggregate ρ_{ag} is estimated as follows [16]:

$$\rho_{ag} = \rho_{dens} \left(\frac{R_{ag}}{a} \right)^{d_f - d} \tag{9}$$

where ρ_{dens} is aggregate substance density in dense packing state (ρ_{dens}=2700 kg/m^3 for carbon [3]), a is lower boundary of aggregate fractal behaviour accepted equal to 15 mnm [17]. For d_f=2.5, d=3 ρ_{ag}≈1300 kg/m^3 will be obtained, that is more than two times smaller than density of filler mass content into value ρ_{ag} and calculate molecular (real) draw ratio λ_{mol}=λ_p of polymeric matrix according to the equation (5). Estimating shear modulus G according to the known equations of high-elasticity classical theory [18], let's calculate Poisson's ratio according to the equation [19]:

$$\nu = \nu_r (1 - \varphi_f) + \varphi_f \nu_f \tag{10}$$

where ν_r and ν_f are values of Poisson's ratio for rubber and filler, accepted equal to 0.50 and 0.25, accordingly.

Further the main equation of high-elasticity fractal theory for σ calculation in its simplified form will be used [9]:

$$\sigma = \frac{E}{1 + 2\nu + 4\nu^3} \left[\lambda_{mol}^{1+2\nu} - 2\nu \lambda_{mol}^{-1-2\nu(1+\nu)} - (1 - 2\nu) \lambda_{mol}^{-2\nu} \right] \tag{11}$$

where elasticity modulus is defined this way [20]:

$$E = 2(1 + \nu) G \tag{12}$$

As follows from the data of Fig. 1, where the comparison of experimental σ^e is made and calculated according to the considered above method σ^T stress of SKI-3 and based on it composites with technical carbon content 20, 30, 40 and 60 mass. % at two testing temperatures (250 and 380 K), good enough correspondence of theory and experiment is obtained, despite arbitrary choice of important parameters number. Two interesting moments should be noted. Firstly, for nonfilled rubber SKI-3 the calculation is made at ν=0.5. Secondly, from the plot of Fig 1 follows, that at small φ_f the values σ^T are somewhat more σ^e and at large φ_f the relation is opposite. This observation can assume an aggregation amplification at filler's content increase, i.e., the R_{ag} rise or aggregates compactness reduction, i.e., d_f decrease.

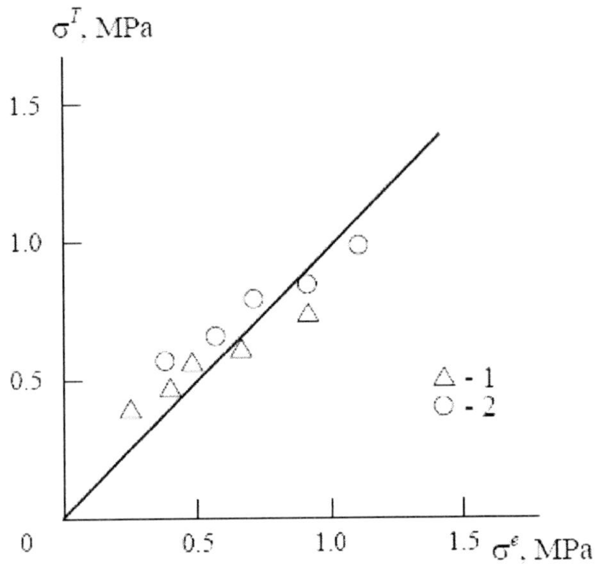

Figure 1. The comparison of experimental σ^e and calculated according to the equation (11) σ^T stress values at temperatures 250 (1) and 380 K (2) for rubber SKI-3, filled by technical carbon

As it is known [18], within the framework of high-elasticity classical theory the value σ is described with the aid of the two following equations:

$$\sigma = G\left(\lambda^2 - \lambda^{-1}\right) \tag{13}$$

$$\sigma = A\left(\lambda - \lambda^{-1/2}\right) \tag{14}$$

where A is material constant.

In Fig. 2 and 3 the dependences σ on generalized stress for studied rubbers, corresponding to the equations (13) and (14), are shown. As can be seen, in case of composites the linearity of these dependences is violated, i.e., at least, the filled rubbers behaviour does not corresponded to high-elasticity classical theory, that is assumed above. Differently speaking, filled rubbers are impossible to consider as ideal, for which internal energy change ΔU is equal to zero in deformation process.

Within the framework of high-elasticity theory the value G in the equation (13) is determined as follows [18]:

$$G = NkT \tag{15}$$

where N is active chains number on rubber volume unit.

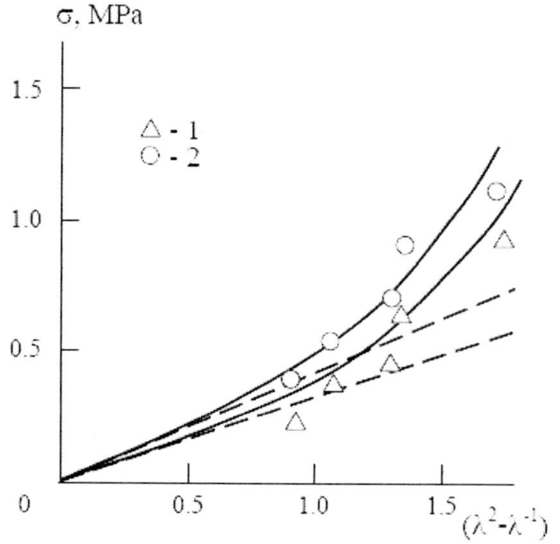

Figure 2. The dependence of stress σ on generalized strain $(\lambda^2-\lambda^{-1})$ at temperatures 250 (1) and 380 K (2) for rubber SKI-3, filled by technical carbon. The shaded lines are shown the dependencies assumed by high-elasticity classical theory

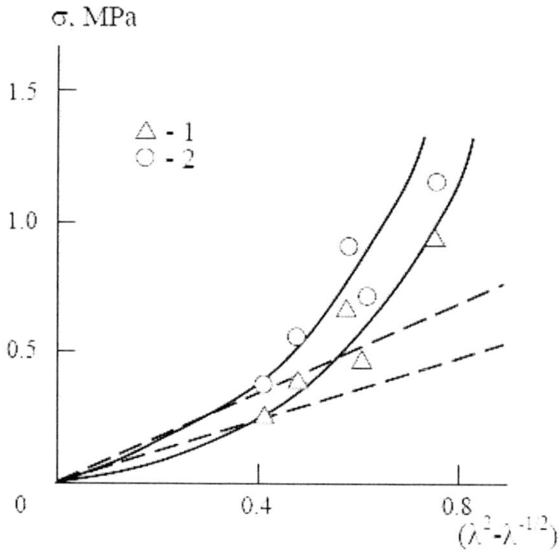

Figure 3. The dependence of stress σ on generalized strain $(\lambda-\lambda^{-1/2})$ at temperatures 250 (1) and 380 K (2) for rubber SKI-3, filled by technical carbon. The shaded lines are shown the dependencies assumed by high-elasticity classical theory

From combination of the equations (13) and (15) follows, that at T=0 K G=0 and σ=0. However, as it is shown in paper [13], it is true only for nonfilled SKI-3, but for filled samples the linear dependence $\sigma(T)$ at T=0 K is extrapolated to nonzero value $\sigma(\sigma_0)$, which is raised at φ_f increase. Besides, from the data of Fig. 2 the deviation value of the dependence $\sigma(\lambda^2-\lambda^{-1})$ Δσ from its linear extrapolation (it is shown in Fig. 2 by shaded line) can be estimated, which is supposed according to the high-elasticity classical theory (see the equation (13)). In a Fig. 4 and 5 the dependences σ_0 and $\Delta\sigma$ on φ_f and $(d\text{-}d_f)$, accordingly are given (for convenience they were linearized by using quadratic form of dependence). As can be seen, in both cases linear correlation passing thourgh coordinates origin for φ_f=0 and d_f=d is obtained, i.e., for nonfilled rubber, which is Euclidean object.

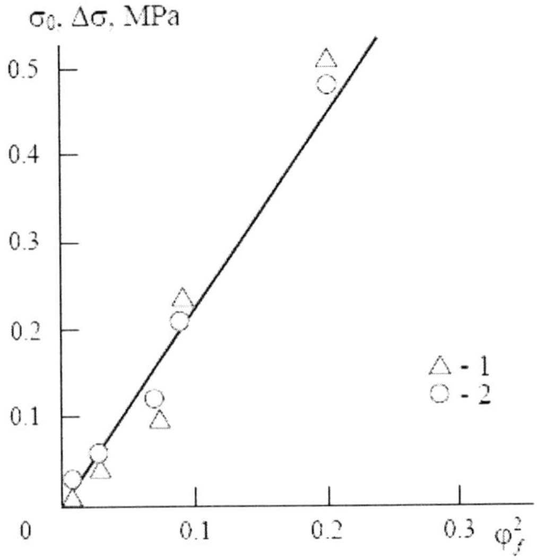

Figure 4. The dependence of deviations from high-elasticity classical theory σ_0 (1) and $\Delta\sigma$ (2) (the explanations give in the text) on filler volumetric degree φ_f for rubber SKI-3, filled by technical carbon

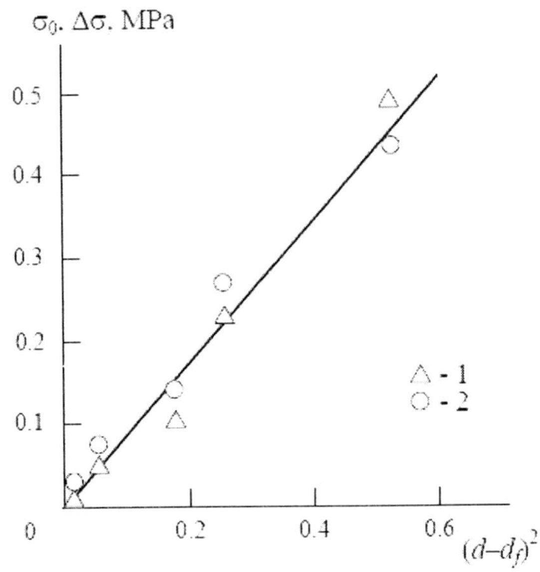

Figure 5. The dependence of deviations from high-elasticity classical theory σ_0 (1) and $\Delta\sigma$ (2) (the explanations give in the text) on Euclidean and fractal dimensions difference $(d\text{–}d_f)$ for rubber SKI-3, filled by technical carbon

CONCLUSION

Therefore, in the present paper three fundamental factors controlling filling processes of nanocomposites with high-elasticity matrix are considered. They are: filler particles aggregation, strain localization and change of structure type from Euclidean object up to fractal. The last postulate is the most physically fundamental, since it demonstrates noncorrectness of description of structure and properties of nanocomposites with high-elasticity matrix within the framework of classical theory of entropic high-elasticity.

REFERENCES

[1] Lipatov, Yu.S. *in Kompozitsionnye Polimernye Materialy.* Kiev, Naukova Dumka, p. 75 (1975).

[2] Aloev, V.Z., Kozlov, G.V., *Physics of Orientational Phenomenas in Polymeric Materials*. Nal'chik, Polygraphservis and T, 2002, 288 p.

[3] Kozlov, G.V., Yanovskii, Yu.G., Lipatov, Yu.S., *Mekhanika Kompozitsionnych Materialov i Konstruktsii*, 9: 398 (2003).

[4] Kozlov, G.V., Lipatov, Yu.S., *Poverchnost'*, 8: 81 (2003).

[5] Feder, F., *Fractaly*. Moscow, Mir, 1991, 256 p.

[6] Shogenov, V.N., Kozlov, G.V., *Fraktal'nye Klastery v Fiziko-Khimii Polimerov.* Nal'chik, Polygraphservis and T, 2002, 268 p.

[7] Watts, M.P.C., Zachariades, A.E., Porter, P.S., *J. Mater. Sci.*, 15: 426 (1980).

[8] Castellani, L., Maestrini, C., *Polymer*, 31: 2278 (1990).

[9] Balankin, A.S., *Pis'ma v ZhTF*, 17: 68 (1991).

[10] Kozlov, G.V., Zaikov, G.E., *Structure of the Polymer Amorphous State*. Utrecht-Boston, Brill Academic Publisher, 2004, 465 p.

[11] Kozlov, G.V., Temiraev, K.B., Shustov, G.B., Mashukov, N.I., *J. Appl. Polymer Sci.*, 85: 1137 (2002).

[12] Balankin, A.S., *Synergetics of Deformable Body*. Moscow, Publisher of Ministry for Defence SSSR, 1991, 404 p.

[13] Yanovskii, Yu.G., Zgaevskii, V.E., *Fizicheskaya Mezomekhanika*, 4: 63 (2001).

[14] Kozlov, G.V., Shogenov, V.N., Mikitaev, A.K., *Inzhenerno-Fizicheskii Zhurnal*, 71: 1012 (1998).

[15] Bobryshev, A.N., Kozomasov, V.N., Babin, L.O., Solomatov, V.I., *Sinergetika Kompozitsionnych Materialov*. Lipetsk, NPO ORIUS, 1994, 154 p.

[16] Brady, R.M., Ball, R.C., *Nature*, 309: 225 (1984).

[17] Avnir, D., Farin, D., Pfeifer, P., *Nature*, 308: 261 (1984).

[18] Bartenev, G.M., Frenkel, S.Ya., *Fizika Polimerov*. Leningrad, Khimiya, 1990, 432 p.

[19] Shustov, G.B., Afaunova, Z.I., Kozlov, G.V., *Vestnik KBGU. Khimicheskie nauki*, 3: 45 (1990).

[20] Kozlov, G.V., Sanditov, D.S., *Anharmonic Effects and Physical-mechanical Properties of polymers.* Novosibirsk, Nauka, 1994, 261 p.

In: Polymers, Polymer Blends, Polymer Composites... ISBN 1-60021-168-2
Eds: A.K. Mikitaev, M.K. Ligidov et al., pp.69-75

Chapter 7

AN INFLUENCE OF MICA SURFACE ON MODEL REACTION OF REETHERIFICATION

Lyubov Kh. Naphadzokova* and Georgi V. Kozlov
Kabardino-Balkarian State University, Nal'chik, Russian Federation

ABSTRACT

The applicability of scaling approach for analysis of mica catalytic activity in model reaction of reetherification is shown. The change of space dimension, in which passing reaction, essentially influences on its intensity. For reaction rate increase is required raising of both Euclidean space dimension and diffusivity of reagents.

Keywords: Polyether, reetherification, catalysis, mica, reaction kinetics, scaling approach.

INTRODUCTION

Composite polymeric materials with improved properties find wide applicability in modern technology. One of the most progressive trends of filled materials making, where the filler defines synthesis kinetics, structure and properties of final product and at the same time performs the role of catalyst, is polymer's synthesis in the presence of inorganic fillers [1, 2].

Saturated complex polyethers, particular polybutylenetherephtalate (PBT), are used as enginiring thermoplasts, having a good thermo- and wear stability, excellent performance. These properties allow also to apply them as a matrix material for polymer composites [3]. One of the perspective ways of effective filler-catalysts searching is kinetic study of reetherification model reaction, performed in the presence of various inorganic compounds. The elucidation on the example of model system of the most effective filler-catalysts number allows to use them for receiving filled PBT and compare catalytic activity of filler and traditional catalysts.

*Correspondence to: Lyubov Kh. Naphadzokova, Kabardino-Balkarian State University, Nal'chik, Russian Federation, Chernyshevski 173, 360004, mailto: lnaph@freemail.ru

The catalysis process reflects a complex interconnection between chemical and geometric parameters of surface. That is why the purpose of the present paper is clarification of influence of filler-catalyst (mica) surface structure on kinetics of reetherification model reaction.

EXPERIMENTAL

The reetherification model reaction kinetics of methylbenzoate by heptanole-1 in mica presence is studied at 443 K. Mica catalytic activity is estimated on the dependence conversion degree-time (*Q-t*) at twentieth multiple of heptanole-1 excess and mica contents 30 mass. % in calculation on the methylbenzoate [2].

The reetherification kinetics is studied on the gas chromatograph "Biokhrom" with using as internal standard diphenyloxide according to the earlier described method [1].

The mica flagopit with polydispersity 0.749 and average probable particles size 0.23×10^{-6} m is used. The initial mica (conditional designation NMM) and also mica chemically modified by sodium hydroxide (SMM) and sulphur acid (AMM) were applied.

RESULTS AND DISCUSSION

For influence analysis of mica surface on reetherification reaction the scaling approach is used [4]. As an example let's consider the reaction in which particles P of a chemical substance diffuse in the medium, containing the randomly located static nonsaturated traps T. By the contact of a particle P with a trap T, the particle disappears. Nonsaturation of a trap means that the reaction P+T→T can repeat itself an infinite number of times. It is usually considered that the concentration of particles and traps is large or the reaction occurs at intersive stirring, and the process can be considered as the classical reaction of the first order. In this case, it is possible to consider that the concentration decay of particles c decreases with time *t* as [4]:

$$c(t) \approx \exp(-At) \tag{1}$$

where A is constant, proportional to the trap concentration.

However, if the concentration of the randomly located traps is small, with necessity exist space areas (reaction medium), practically free from traps. The particles getting into these areas can reach the traps only during a rather long period of time and, hence, the decrease of their number in the reaction course will be slower. The formal analysis of this problem shows that the concentration of particles falls down under the law [4]:

$$c(t) \sim \exp\left(-Bt^{d/(d+2)}\right) \tag{2}$$

being dependent on the space dimension d (B is constant).

It is necessary to mark, that the singular dependence on time in the equation (2) appears simultaneously with large-scale fluctuations (inhomogeneity) of traps density. If the traps can move, their mobility averages the influence of spatial heterogeneity, so the assumptions resulting to (1) will be carried out better. It was shown, that in this case the concentration of the particles drops under the combined law [4]:

$$c(t) \sim \exp(-At)\exp\left(-Bt^{d/(d+2)}\right) \qquad (3)$$

where A is proportional to the traps diffusivity.

Since the reetherification reaction elapses with intensive stirring due to both passing of inert gas through reactive medium and mechanical stirring ("scavenger" reaction [5]), then for theoretical description of reagents concentration decay $c(t)(=1-Q)$ with time t we use the relationship (3). In Fig. 1-3 the fit of theory and experiment for reetherification reactions without mica (Fig. 1), in the presence of NMM (Fig. 2) and AMM (Fig. 3) is shown. The data for SMM are analogous to the shown in Fig. 2 results for NMM and therefore are not shown. Two space dimensions d are used, in which the reaction elapses: d=2 and d=3. As can be see reetherification reaction without mica is approximately equally described by both cases, but better correspondence is reached by using d=3. For reaction in the presence of NMM (and SMM) the space transition from d=3 to d=2 is observed, i.e., such transition assumes elapsing of reetherification reaction on the flat surface (the mica surface dimension d_{sp}=2 [6]) of filler-catalyst. And at last in case of AMM the space dimension becomes again equal to 3.

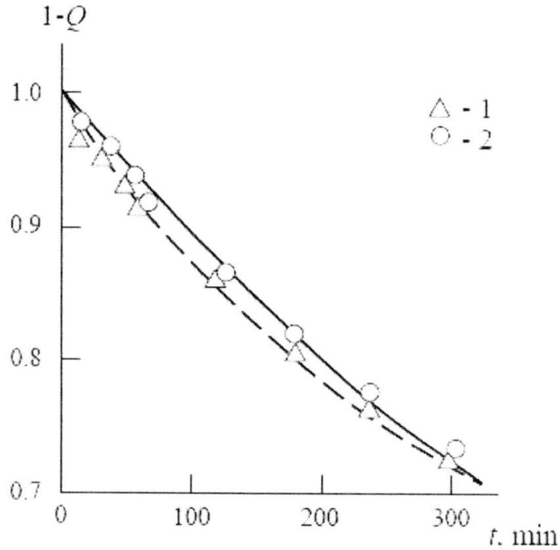

Figure 1. The dependences (1-Q) on t, corresponding to the relationship (3), for reetherification reaction without mica at d=2 (1) and d=3 (2). Solid line is experimental data

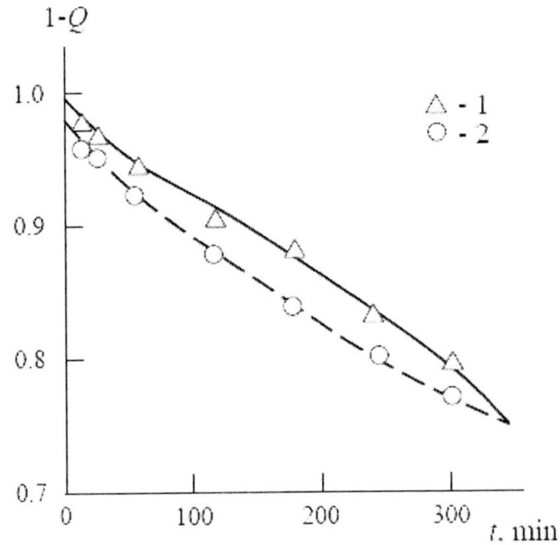

Figure 2. The dependences (1-Q) on t, corresponding to the relationship (3), for reetherification reaction in presence of NMM. The notation is the same, as in a Fig. 1

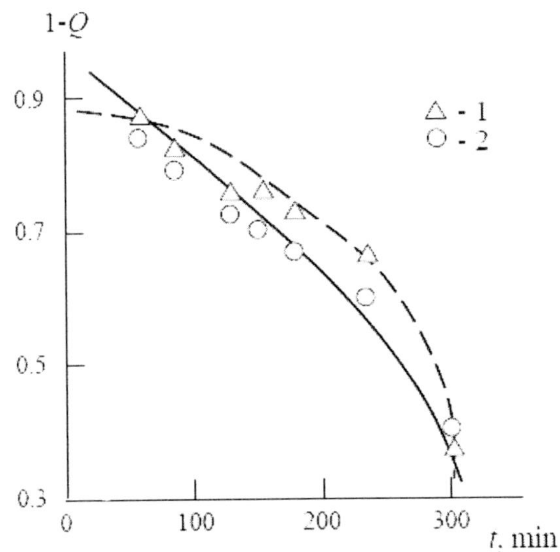

Figure 3. The dependences (1-Q) on t, corresponding to the relationship (3), for reetherification reaction in presence of AMM. The notation is the same, as in a Fig. 1

Let's consider physical principles of mentioned dimension changes and, hence, space type in which considered reaction elapses. The value d_{sp} for three mica types can be estimated with the aid of the relationship [7]:

$$k_{2KT} \sim t^{(1-d_{sp})/2} \tag{4}$$

where k_{2KT} is catalysis rate constant of the second order accepting according to the data of paper [2]. The constant in the relationship (4) can be determined at condition d_{sp}=2.0 for NMM [6].

In table 1 the relationship (3) constants are accepted at which using the best correspondence with experiment (A, B and d) is obtained and the values d_{sp} are also estimated according to the equation (4). As can be seen, mica processing by sodium hydroxide practically does not change the value d_{sp} for SMM in comparison with the initial mica NMM, whereas processing by sulphur acid results to the fact, that AMM surface becomes fractal object with dimension d_{sp}=2.23. If the values d_{sp} for NMM and SMM characterize the essence of Euclidean surface and, hence, are considered in d=2, then for AMM fractal surface with d_{sp}>2 can be considered only in Euclidean space with d=3, that is observed according to the data of Fig. 3 and Table 1. The increase d_{sp} for AMM in comparison with similar value for NMM and SMM completely co-ordinates with specific surface raising S_u for the first (S_u≈21.3 and 18.4 m^2/g, accordingly [2]).

The space dimension change results to variation of reactive medium effective volume V_{ef}, in which reaction can elapse. The value V_{ef} is determined as that [8]:

$$V_{ef} = L^{d_{ef}} \varepsilon^{3-d_{ef}} \tag{5}$$

where L is upper limit of fractal behaviour, ε is measurement scale, d_{ef} is effective medium dimension.

Table 1. The characteristics of filler-catalyst (mica), applied in model reaction of reetherification

Mica	d	$A\times10^4$	$B\times10^3$	d_{sp}	V_{ef}, relative units
Is not	3	5.2	5.0	-	8
NMM	2	5.2	4.0	2.0	2
SMM	2	4.8	3.9	1.964	2
AMM	3	10.0	10.0	2.230	8

The estimation V_{ef} according to the equation (5) at arbitrary values L=2, ε=0.5 and condition d_{ef}=d shows essential distinction of effective volume for considered reactions (table 1). As it has been noted above, the constant A in the relationship (3) is proportional to traps diffusivity [4, 5]. It is necessary to wait, that the increase both V_{ef} and A (traps diffusivity) results to the conversion degree Q raising of reetherification reaction. In Fig. 4 the dependence of value Q at t=300 min. on product AV_{ef} (in relative units) for studied reactions confirmed this assumption is shown. The pointed dependence is linear, i.e., factors A and V_{ef} have equal influence degree on value Q, and passes through coordinates origin, i.e., obvious condition is observed: at A=0 or V_{ef}=0 Q=0.

The expounded above results allow to explain the kinetics of reetherification model reaction. In Fig. 5 kinetic curves Q-t for four studied reactions are shown. They are all linear, i.e., reetherification reaction elapses in Euclidean space with dimension d=2 or d=3 (see table 1). The reaction without mica occupies intermediate position because of a larger dimension d, but relatively small diffusivity.

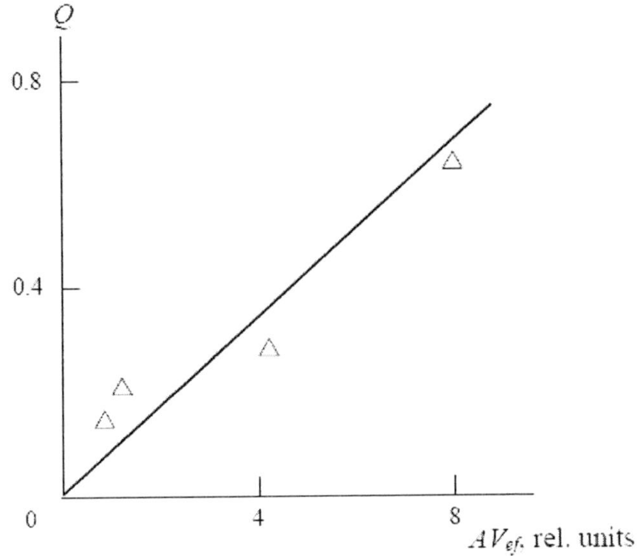

Figure 4. The dependence of conversion degree Q at t=300 min on product AV_{ef} for reetherification reactions

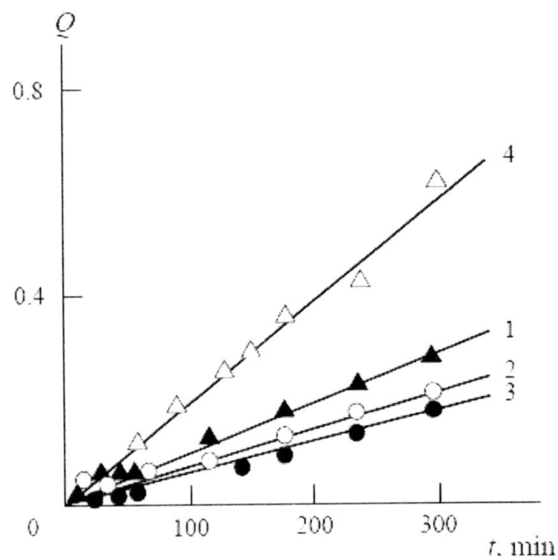

Figure 5. Kinetic curves conversion degree-time (Q-t) for reetherification reactions without mica (1) and in presence of NMM (2), SMM (3), AMM (4)

Introduction in reactive medium NMM and SMM results to the decrease of both d and A, that reduces the reaction intensity. And at last, for reetherification reaction in the presence of AMM are obtained the largest values of both d (d=3) and A (diffusivity), by virtue of which this reaction elapses much more rapidly than the rest. Also let's note, that coefficient B in the relationship (3) is proportional to reactive sites number of reagents [5] and its value is the largest for reetherification reaction in the presence of AMM (Table 1).

CONCLUSION

Therefore, the results of present paper have shown the applicability of scaling approach for analysis of mica catalytic activity in model reaction of reetherification. The change of space dimension, in which the reaction passes, essentially influences on its intensity. For reaction rate increase is required raising of both Euclidean space dimension and diffusivity of reagents.

REFERENCES

[1] Vasnev, V.A., Naphadzokova, L.Kh., Tarasov, A.I., Vinogradova, S.V. Lependina, O.L., *Vysokomol. Soed A*, 42: 2065 (2003).

[2] Naphadzokova, L.Kh., Vasnev, V.I., Tarasov, A.I., *Plast. Massy*, 3: 39 (2001).

[3] Mikitaev, A.K., Kalajyan, A.A., Lednev, O.B., Mikitaev, M.A., *Plast. Massy*, 12: 45 (2004).

[4] Djordjevič, Z.B., *in Fractals in Physics*. Moscow, Mir, 1988, p. 581.

[5] Redner, S., Kang, K., *J. Phys. A*, 17: L451 (1984).

[6] Van Damme, H., Fripiat, J.J., *J. Chem. Phys.*, 82: 2785 (1985).

[7] Kozlov, G.V., Shustov, G.B., Zaikov, G.E., *J. Balkan Tribolog. Assoc.*, 9: 467 (2003).

[8] Van Damme, H., Levitz, P., Bergaya, F., Alcover, J.F., Gatineau, L., Fripiat, J.J., *J. Chem. Phys.*, 85: 616 (1986).

In: Polymers, Polymer Blends, Polymer Composites... ISBN 1-60021-168-2
Eds: A.K. Mikitaev, M.K. Ligidov et al., pp.77-81

Chapter 8

THE INTERRELATION OF ELASTICITY MODULUS AND AMORPHOUS CHAIN'S TIGHTNESS FOR NANOCOMPOSITES BASED ON THE POLYPROPYLENE

Georgi V. Kozlov*[1], *Ahmed Kh. Malamatov*[1*], *Eugeni M. Antipov*[2] and *Abdulah K. Mikitaev*[3]

[1]Kabardino-Balkarian State University, Nal'chik, Russia
[2]Topchiev Institute of Petrochemical Synthesis of Russian Academy of Sciences, 119991 Leninskii pr. 29, Moscow, Russia
[3]State Scientific Institution "Compositecenter",125047 Miusskii Square 9, Moscow, Russia

ABSTRACT

The interrelation of elasticity modulus and amorphous chain's tightness characterized by fractal dimension of chain part between its fixation points for nanocomposites based on the polypropylene is shown. This assumes the polymeric matrix structure change in comparison with initial polymer: the role of densely-packed regions for it is played by interphase areas. An offered fractal model allows estimation of elasticity modulus limiting values.

Keywords: Nanocomposite; polypropylene; elasticity modulus; amorphous chain tightness; fractal analysis.

[*]Correspondence to: Ahmed Kh. Malamatov, Kabardino-Balkarian State University, Nal'chik, Russia, Chernyshevski 173, 360004, mailto: deanchem@ns.kbsu.ru

INTRODUCTION

At present time it is assumed [1], that the elasticity modulus E of semi-crystalline polymers is controled by structure (topology) of noncrystalline areas. The authors [2] were offered the model, according which value E increases at amplification of chains tightness in amorphous areas (amorphous chains) of the mentioned polymers. This model was applied succesfully for description of elasticity modulus behaviour of compositions based on the high density polyethylene modified by highly dispersed mixture Fe/FeO [1]. However, the degree of amorphous chains tightness can also be experessed within the framework of fractal analysis using for this dimension D of chain part between its fixation point (physical entanglements, chemical cross-linking nodes and so on) [3]. The value D is varied in limits $1<D\leq2$ and at D=1 the chain is fully streched between its fixation points (mobility of this chain part is freezed). At D=2 the chain has maximally possible mobility which is typical for rubber state of polymers [3]. This allows enough simple and adequate description of amorphous chains tightness degree with the aid of dimension D. The purpose of present paper is to give the development of fractal analoque of model [2] and with its aid to receive the description of behaviour of elasticity modulus of nanocomposites based on the polypropylene.

EXPERIMENTAL

The data of paper [4] for nanocomposites based on industrial isotactic polypropylene (PP) of manufacture Shell Co. is used. As a filler natural clay Na^+-montmorillonite is used modified by four modificators, with volume contents φ_f=0.025, 0.050 and 0.10. The more detailed description of nanocomposites in paper [4] is cited. Specimens for mechanical testing in the form of dog-bone with base lenght 10 mm and thichness 0.2 mm were produced by pressing from the melt at 473 K. The tension testing was made at temperatures 293 and 373 K and strain rate 8.3×10^{-3} s^{-1} [4].

RESULTS AND DISCUSSION

As it is known [4], Na^+-montmorillonite is layered silicate the plates of which have smaller thichness than their width and length. Therefore it is assumed, that the structure of nanocomposite having that filler besides actually Na^+-montmorillonite has crystalline regions, amorphous (rubber) phase and interfacial areas. In this case reinforcement degree E_c/E_m can be written in the following form [5]

$$\frac{E_c}{E_m} = \left[1 + 11(\varphi_f + \varphi_{\text{int}})^{1.7}\right] \tag{1}$$

where E_c and E_m are elasticity modulus of nanocomposite and matrix polymer accordingly, φ_{int} – relative fraction of interfacial areas. The equation (1) allows to determine the φ_{int} value on the base of experimental values E_c, E_m and φ_f.

The calculation of dimension D was made according to following method. First ofall fractal (Hausdorff) dimension of nanocomposite structure d_f was determined according to the equation [3]

$$d_f = (d-1)(1+\nu) \tag{2}$$

where d is dimension of Euclidean space, in which the fractal is considered (obviously, in our case d=3), ν is Poisson's ratio estimated on the mechanical testing results with the aid of relationship [1]

$$\frac{\sigma_Y}{E} = \frac{1-2\nu}{6(1+\nu)} \tag{3}$$

where σ_Y is yielding stress.

The value of characteristic ratio C_∞, which is the indicator of polymeric chain statistical flexibility, is determined according to the following equation [3]

$$C_\infty = \frac{2d_f}{d(d-1)(d-d_f)} + \frac{4}{3} \tag{4}$$

Orthorhombic crystallinity degree K of polymer is connected with parameter C_∞ by the equation [6]

$$K = 0.32C_\infty^{1/3} \tag{5}$$

Now the value of fractal dimension D can be calculated with the aid of the following equation [7]

$$\frac{2(1-K)}{\varphi_{\text{int}}} = C_\infty^D \tag{6}$$

In a Fig. 1 the dependences of elasticity modulus E_c on dimension D at two mentioned above testing temperatures T for nanocomposites based on the PP are shown. Just as it was expected, linear increase of E_c at decrease D or amplification of amorphous chains tightness is observed. Two various straight lines for two testing temperatures are obtained, according which the maximally reached values $E_c^{\max}$ can be determined: at T=293 K $E_c^{\max}$=1.16 GPa, at T=373 K $E_c^{\max}$=0.53 GPa. It should be noted, that one of the advantages of fractal analysis is availability of precisely determined dimension limits (for example, $1<D\leq 2$ [7]), that allows estimation of limiting values of any property. The data of Fig. 1 allow to receive analytical relationship between E_c and D in the following form

$$E_c = E_c^{\max} - E_c^{\max}(D-1) \text{ or} \tag{7}$$

$$E_c = E_c^{\max}(2-D) \tag{8}$$

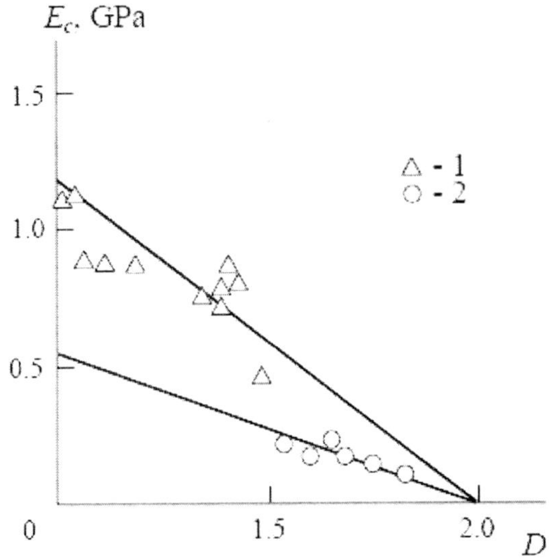

Figure. 1. The elasticity modulus E_c vs. fractal dimension of chain part between interfacial areas D at testing temperatures 293 (1) and 373 K (2) for nanocomposites based on the polypropylene

CONCLUSION

Therefore, the results of the present paper showed interrelation of elasticity modulus and amorphous chains tightness characterized by fractal dimension of chain part between its fixation points for nanocomposites based on the semi-crystalline polypropylene. This assumes the polymeric matrix structure change in comparison with initial polymer: the role of densely-packed regions for it is played by interfacial areas. An offered fractal model allows estimation of elasticity modulus limiting values.

REFERENCES

[1] Kozlov, G.V., Sanditov, D.S., *Anharmonic Effects and Physical-mechanical Properties of polymers.* Novosibirsk, Nauka, 1994, 261 p.

[2] Krigbaum, W.R., Roe, R.-G., Smith, K.J., *Polymer*, 5: 533 (1964).

[3] Novikov, V.U., Kozlov, G.V., *Polymer's Fracture Analysis within the Framework of Fractals Conception.* Moscow, MSOU Publisher, 2001, 136 p.

[4] Antipov, E.M., Barannikov, A.A., Gerasin, V.A., Shklyaruk, B.F., Tsamalashvili, L.A., Fisher, H.R., Razumovskaya, I.V., *Vysokomolek. Soed.* A, 45: 1885 (2003).

[5] Kozlov, G.V., Burya, A.I., Aloev, V.Z., Grineva, L.G., in *Mechaniks and Management Processes*, Ekaterinburg, p. 97 (2004).
[6] Aloev, V.Z., Kozlov, G.V., *Physics of Orientational Phenomenas in Polymeric Materials*. Nal'chik, Polygraphservis and T, 2002, 288 p.
[7] Kozlov, G.V., Zaikov, G.E., *Structure of the Polymer Amorphous State*. Utrecht-Boston, Brill Academic Publisher, 2004, 465 p.

In: Polymers, Polymer Blends, Polymer Composites… ISBN 1-60021-168-2
Eds: A.K. Mikitaev, M.K. Ligidov et al., pp.83-88

Chapter 9

STRUCTURE FORMATION OF POLYMER NANOCOMPOSITES BASED ON POLYPROPYLENE

Ahmed Kh. Malamatov[1]*, Georgi V. Kozlov[1] and Eugeni M. Antipov[2]
[1]Kabardino-Balkarian State University, 360004 Chernyshevski st. 173, Nal'chik, Russia
[2]Topchiev Institute of Petrochemical Synthesis of Russian Academy of Sciences, 119991 Leninskii pr. 29, Moscow, Russia

ABSTRACT

It is shown, that the structure formation of nanocomposites based on the polypropylene takes in Euclidean space with dimension d=3. Within the framework of the offered for these material structural model is shown identity of local order regions, interfacial areas and crystalline β-phase. This model allows quantitative description of nanocomposites properties.

Keywords: Nanocomposite; polypropylene; structure; Euclidean space; local order; crystalline β-phase.

INTRODUCTION

The introduction of disperce filler in polymeric matrix changes significantly properties of the last [1]. This fact is well known, but there are definite difficulties with its quantitative estimation. For particulate-filled composites polyhydroxiether-graphite (PHE-Gr) was found the structure fractal dimension d_f increase from 2.6 to 2.8 within the volume content of filler interval φ_f=0-0.09. This structure change is accompanied by corresponding increase of composites mechanical characteristics (elasticity modulus, yielding stress, fracture stress and so on) [2]. The mentioned change of composites PHE-Gr structure is due to its formation in fractal space, which creates a filler network of particles (particle aggregates) [3,4]. As studies of mechanical properties of nanocomposites based on the polypropylene (PP), filled by modified natural clay (NC) Na^{+}-montmorillonite in interval φ_f=0.025-0.10, show that for them elasticity modulus raising is not observed and yielding stress and strength are even a little decreased [5]. Proceeded from these differences of mechanical properties behaviour of composites PHE-Gr and nanocomposites based on the PP, presents interest of the study of structural changes, due to introduction of layered filler and determines mechanical behaviour of nanocomposites, that is the purpose of present paper.

EXPERIMENTAL

The data of paper [5] for nanocomposites based on the isotactic industrial PP (Shell Co.), filled by Na^{+}-montmorillonite with last contents φ_f=0.025, 0.050 and 0.10 were used. As modificators: dioctadecyldimethyl ammonium bromide (DODAB) plus block-copolymer polyethyleneoxide-polyethylene (PEO-PE) (conventional sign of nanocomposite PP-NC-1); PEO-PE (PP-NC-2); DODAB plus PEO-PE with isobutylene (PP-NC-3); PEO-PE plus isobutylene (PP-NC-4) are used. The detailed description of specimens preparation methodics is cited in the paper [5].

The mechanical tests were made on film specimens with length 10 mm, width 3 mm and thickness 0.2 mm, preparated by pressing from melt at 473 K. The tests were made at temperature 293 K and strain rate 8.3×10^{-3} s^{-1} [5].

The experimental values d_f(d_f^e) were determined according to the equation [6]

$$d_f^e = (d-1)(1+\nu) \tag{1}$$

where d is dimension of Euclidean space, in which a fractal is considered (obviously, in our case d=3), ν is Poisson's ratio, the value of which can be estimated on mechanical testing results with the aid of the relationship [7]

$$\frac{\sigma_Y}{E} = \frac{1-2\nu}{6(1+\nu)} \tag{2}$$

where σ_Y is yielding stress, E is elasticity modulus.

RESULTS AND DISCUSSION

As it is mentioned above, the change d_f at increase φ_f for composites PHE-Gr is due to formation of polymeric matrix structure in fractal space with dimension D_{lat} [3,4]. The value can be determined with the aid of the following equation [8]

*Correspondence to: Ahmed Kh. Malamatov, Kabardino-Balkarian State University, Nal'chik, Russia, Chernyshevski 173, 360004, mailto: deanchem@ns.kbsu.ru

$$\nu_F = \frac{3}{2 + D_{lat}} \tag{3}$$

where ν_F is Flory exponent connected with dimension d_f in the linear polymers case by the following relationship [9]

$$\nu_F = \frac{1.5}{d_f} \tag{4}$$

The calculation of value d_f^e according to the equation (1) shows this dimension variation within the limits 2.678-2.739 with average value 2.709. The estimation D_{lat} according to the equation (3) give in this case the value ~ 3.42, that is does not correspond to condition $D_{lat} \leq 3$, which is common for all objects in three-dimensional Euclidean space [6]. Therefore, it is necessary to assume $D_{lat}=3$ and value d_f in this case is equal to 2.5 according to the equation (3). This value d_f is excellently coordinated with calculation according to the equation (1) for initial PP (d_f^e=2.518). Thus, the cited estimations allow to assume, that the formation of structure of initial PP and polymeric matrix of nanocomposites on its basis is realized in the three-dimensional Euclidean space, and increase d_f^e for nanocomposites in comparison with PP is due to introduction of the filler. It is necessary to mark, that at condition $D_{lat}<3$ (structure formation in fractal space) the decrease d_f in comparison with $D_{lat}=3$ for nanocomposites in comparison with initial PP must take place, that contradicts to experimental data.

Let's consider physical base of the mentioned increase d_f^e. As it is shown in paper [5], the introduction of layered filler (Na^+-montmorillonite) in PP results to the formation of crystalline β-phase, the crystallites of which have primary orientation inseparably connected with flat texture of siliceous layers, on which they are crystallized. This allows to consider mentioned crystallites both as local order regions, and as interfacial areas at the same time. Let's perform quantitative estimation of these regions contents at both of their definitions. The relative fraction of local order regions (clusters) φ_{cl} for semi-crystalline polymers can be calculated according to the following percolation relationship [9]

$$\varphi_{cl} = 0.03(1-K)(T_m - T)^{0.55} \tag{5}$$

where K is orthorhombic crystallinity degree, T_m and T are temperature of melting and testing accordingly.

For PP $T_m \approx 440$ K [10] and value K can be determined according to the equation [11]

$$K = 0.32 C_\infty^{1/3} \tag{6}$$

where C_∞ is characteristic ratio, which is an indicator of polymer chain statistical flexibility [12] and connected with dimension d_f by the relationship [9]

$$C_\infty = \frac{2d_f}{d(d-1)(d-d_f)} + \frac{4}{3} \quad (7)$$

The relative fraction of interfacial areas φ_{inf} was calculated with the aid of the equation [13]

$$\frac{E_c}{E_m} = 1 + 11(\varphi_f + \varphi_{inf}) \quad (8)$$

where E_c and E_m are elasticity modulus of composite and matrix polymer, accordingly.

The connection between dimension d_f and crystallinity degree is defined by the following empirical approximation [11]

$$d_f = 2 + K \quad (9)$$

It is obvious, that for the studied nanocomposites with appreciation of crystalline β-phase formation the following version of the equation (9) should be written

$$d_f = 2 + K + \varphi_{inf} \quad (10)$$

According to the equation (10) the formation of nanocomposites structure in fractal space, defining decrease d_f, must decrease value K, that also does not correspond to the experiment.

In Fig. 1 the comparison of dimensions d_f^e and d_f^T calculated according to the equations (1) and (10), accordingly, is shown. As can be seen, between these values d_f the good correspondence is received (average discrepancy by fractional part of dimensions, which performs main information about structure, is equal to 4.6 %). This allows to assume, that the structure of nanocomposites based on the PP consists of the following structural regions: orthorhombic crystallites, amorphous phase, filler and crystalline β-phase, forming interfacial areas. Let's note, that the calculation φ_{inf} according to the equation (8) and nanocomposites crystallinity degree rise in comparison with initial PP [5] give close values: ~ 0.15-0.20.

Another method of calculation of theoretical value d_f(d_f^T) gives the following equation [9]

$$d_f^T = 3 - 6\left(\frac{\varphi_{cl}}{SC_\infty}\right)^{1/2} \quad (11)$$

where S is cross-section area of macromolecule, which is equal to 34.27 Å^2 for PP [14].

In Fig. 2 the dependence d_f^T on φ_f calculated according to the equation (4) for PP and according to the equation (11) for nanocomposites based on the PP at φ_{cl} estimated according to the equation (5) is shown. The comparison with experimental data for four groups of nanocomposites and initial PP shows correctness of this estimation.

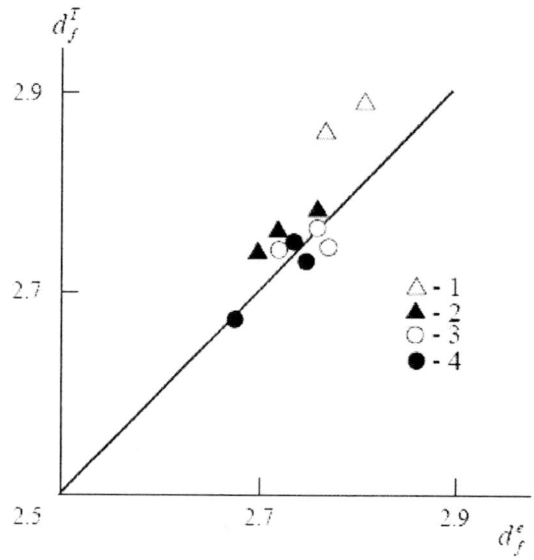

Figure 1. The comparison of experimental d_f^e and calculated according to the equation (10) d_f^T values of structure fractal dimension for nanocomposites PP-NC-1 (1), PP-NC-2 (2), PP-NC-3 (3) and PP-NC-4 (4)

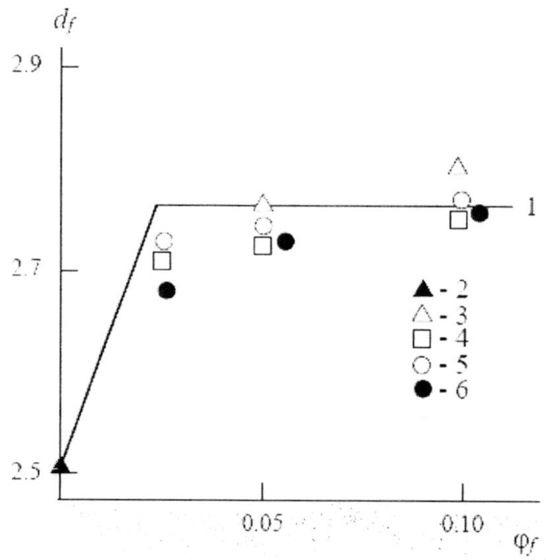

Figure 2. The comparison of calculated according to the equations (4) and (11) (1) and experimental (2-6) the dependences of fractal dimension d_f on filler contents φ_f for PP (2), PP-NC-1 (3), PP-NC-2 (4), PP-NC-3 (5) and PP-NC-4 (6)

CONCLUSION

Therefore, the results of the present paper showed, that the structure formation of nanocomposites based on the polypropylene takes place in Euclidean space with dimension d=3. Within the framework of offered for these materials structural model is shown identity of local order regions, interfacial areas and crystalline β-phase. This model allows quantitative description of nanocomposites properties.

REFERENCES

[1] Lipatov, Yu.S., *Physical-mechanical Principles of Polymer Filling*. Moscow, Khimiya, 1991, 259 p.

[2] Novikov, V.U., Kozlov, G.V., *Mekhanika Kompozitn. Mater.*, 35: 269 (1999).

[3] Kozlov, G.V., Yanovskii, Yu.G., Lipatov, Yu.S., *Mekhanika Kompozitsion. Mater. and Konstruk.*, 8: 467 (2002).

[4] Kozlov, G.V., Lipatov, Yu.S., *Mekhanika Kompozitn. Mater.*, 40: 827 (2004).

[5] Antipov, E.M., Barannikov, A.A., Gerasin, V.A., Shklyaruk, B.F., Tsamalashvili, L.A., Fisher, H.R., Razumovskaya, I.V., *Vysokomol. Soed* A, 45: 1885 (2003).

[6] Balankin, A.S., *Synergetics of Deformable Body*. Moscow, Publisher of Ministry for Defence SSSR, 1991, 404 p.

[7] Kozlov, G.V., Sanditov, D.S., *Anharmonic Effects and Physical-mechanical Properties of polymers*. Novosibirsk, Nauka, 1994, 261 p.

[8] Vannimenus, J., *Physica D*, 38: 351 (1989).

[9] Kozlov, G.V., Zaikov, G.E., *Structure of the Polymer Amorphous State*. Utrecht-Boston, Brill Academic Publisher, 2004, 465 p.

[10] Kalinchev, E.L., Sakovtseva, M.B., *Properties and Processing of Thermoplasts*. Leningrad, Khimiya, 1983, 288 p.

[11] Aloev, V.Z., Kozlov, G.V., *Physics of Orientational Phenomenas in Polymeric Materials*. Nal'chik, Polygraphservis and T, 2002, 288 p.

[12] Budtov, V.P., *Physical Chemistry of Polymer Solutions*. Sankt-Peterburg, Khimiya, 1992, 384 p.

[13] Kozlov, G.V., Burya, A.I., Aloev, V.Z., Grineva, L.G., in *Mechaniks and Management Processes*, Volume 1, p. 97, Ekaterinburg (2004).

[14] Aharoni, S.M., *Macromolecules*, 18: 2624 (1985).

In: Polymers, Polymer Blends, Polymer Composites… ISBN 1-60021-168-2
Eds: A.K. Mikitaev, M.K. Ligidov et al., pp.89-92

Chapter 10

SYNTHESIS AND STUDY OF PROPERTIES OF AROMATIC POLYETHER–IMIDES ON THE BASIS OF DERIVATIVES OF CHLORAL AND DDT WITH USE OF POLYNITROREPLACEMENT PROCESSES

***R. M. Kumykov*[1*], *M. T. Bezhdugova*[1], *A. K. Ittiev*[1], *A. K. Mikitaev*[2] and *A. L. Rusanov*[2]**

[1]Kabardino–Balkarian State Academy of Agriculture, 360004, Nalchik, Tarchokov Street, 1a, Russia

[2]A.N. Nesmejanov Institute of Element – Organic Compounds of Russian Academy of Science, 117813, Moscow, Vavilov Street, 28, Russia

ABSTRACT

The new aromatic polyether–imides containing «hinge» groupings and six–member imide cycles in polymeric macromolecules are obtained with help of polynitroreplacement reaction. It is shown, that polyether–imides soluble in organic solvent and possessing the greater chemical, thermo– and fire– resistance in comparison with the similar systems containing five–member imide cycles in polymeric macromolecules have been obtained at use of sodium salts of aromatic bisphenols and of dinitronaphtalimides (derivatives of chloral and DDT) as co–monomers.

Keywords: polyether–imides, polynitroreplacement, dinitrinaftalimides, dichloroethylene, carbonyl.

*Correspondence to: Kabardino–Balkarian state academy of agriculture, 360004, Nalchik, Tarchokov street, 1a, Russia. mailto: beevaues@front.ru

Introduction

According to works [1–5], nitrogroupings activated by two carbonyls participate in reactions of nucleophilic replacements; in particular, the activation of nitrogroupings by two carbonyls contained in cyclic imides is effective one [4; 6].

Besides the replacing of five–member imide fragments on six–member ones leads to the formation of soluble in organic solvents [7–9] polymers with higher chemical, thermo– and fire resistance.

In view of above–stated, the reaction of interaction between aromatic bisphenols, containing dichloroethylene, carbonyl and methylene «hinge» groupings, and dinitronaphtalimides, being also derivatives of chloral and DDT, was carried out.

Experimental

The reaction was kept in soft temperature conditions (50 – 70 °C) without water (because N–replaced nitronaphtalimides are easily deactivated as a result of hydrolysis and disclosing of a cycle) in dimethyl sulfoxide environment during 2 hours. Reaction resulted in the appearance of final products due to the scheme:

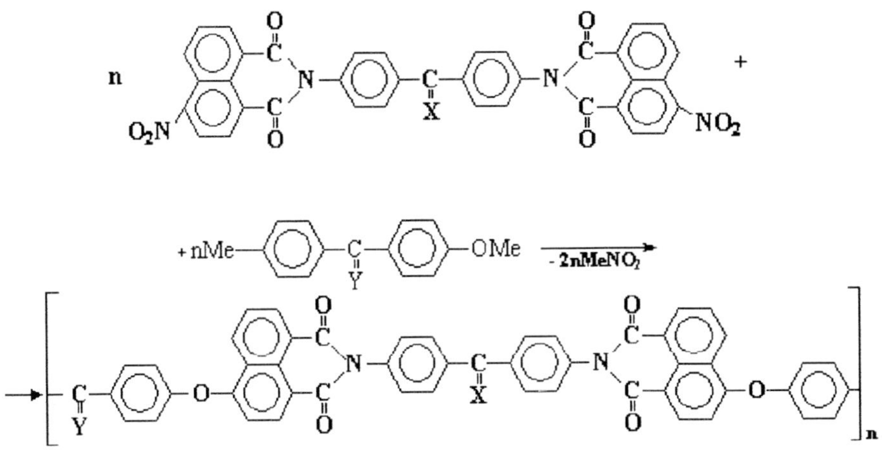

where

$$X = -\underset{\underset{O}{\|}}{C}-,\ -CH_2-;\quad Y = -\underset{\underset{CCl_2}{\|}}{C}-,\ -\underset{\underset{O}{\|}}{C}-,\ -CH_2-.$$

Conditions of synthesis and some characteristics of polyetherimides obtained are listed in table. 1.

Sodium phenoxide was added into reactionary mixes for replacing trailer nitrogroupings and to increase the thermostability in all cases after the end of reaction of synthesis of polyetherimide.

Table 1. Conditions of synthesis and some characteristics of polyetherimides of general formula

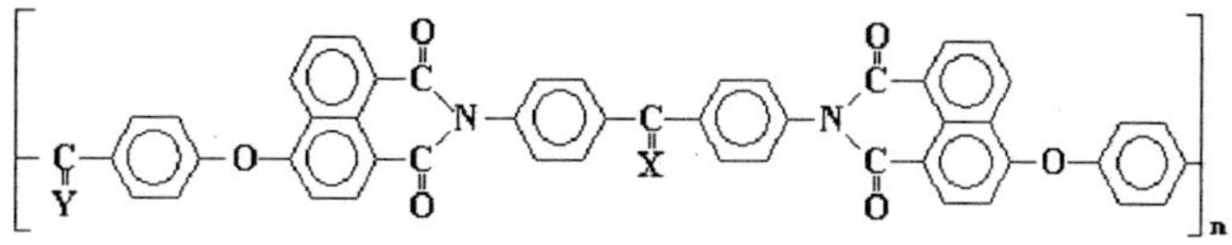

Compound	X	Y	η_{str}, dl/g	T_{sample}, °C	$T_{10\%}$, °C	KI	Output, %	Conditions of synthesis		
								Solvent	Temperature, °C	Duration, hours
1	—C(=O)—	—C(=CCl$_2$)—	0,47	230–240	420	58	85	DMSO / toluene	60	2
2	–C(=O)–	–C(=O)–	0,52	272–281	480	32	91	DMSO / toluene	70	2
3	–C(=O)–	–CH$_2$–	0,60	263–272	450	31	95	DMSO / toluene	60	2
4	–CH$_2$–	–C(=CCl$_2$)–	0,47	235–243	410	58	90	DMSO / toluene	60	2
5	–CH$_2$–	–CH$_2$–	0,55	267–275	420	31,8	87	DMSO / toluene	60	2
6	–CH$_2$–	–C(=O)–	0,6	263–273	450	32,6	94	DMSO / toluene	70	2

The structure of synthesized polyetherimides has been confirmed with the data of the IR–spectral analysis. Particularly, IR–spectra of polyetherimides based on 1,1–dichlor–2,2–di(p–oxiphenyl)ethylene contain maxima of absorption in the range from 840 up to 980 sm^{-1} attributed to 1,1–dichlorethylene groupings [10]. All spectra of polyethernaphtalimides contain doublets at 1738–1745 and 1780–1785 sm^{-1}, characteristic for carbonyl groups of six–member anhydride cycle [10]. Besides the strips of absorption caused by presence of «bridge» groupings occur in IR–spectra: 1260 sm^{-1} (–O–) and 1645–1680 sm^{-1} (–CO–).

The temperatures of a softening of polymers, according to the TMA data, are 230–275 °C, and temperatures of the beginning of decomposition (10 % of weight loss, according to the TGA data) are 410–480 °C.

The feature of obtained polyetherimides is the significant distinction between temperatures of glassing and decomposition of these polymers.

Synthesized polyetherimides, containing dichloroethylene groupings, possess high fire resistance (KI = 58) and good solubility in organic solvents (N–MP, DMFA, DMAA), tetrachloroethane, dichloroethane and chloroform.

Synthesis of Polyetherimides

The NaOH (of 0,04M in 25 ml of H_2O) was poured into the reactionary flask (supplied with the mixer, the Dyne – Stark trap, the return refrigerator and barbator for inert gas) of capacity 1,5 l in where loaded bisphenol (of 0,02 M) and 390 ml of DMSO and 210 ml of toluene were constantly mixed at temperature 80 °C. Then temperature was raised up to 150 °C and water was driven away. After the thorough removal of water traces the temperature of reactionary mass was reduced downto 60–70 °C and dinitrophtalimide (of 0,02 M) was added. Reaction was carried out at this temperature within 2 hours. Then reactionary mass was diluted by DMSO and besieged into the acidified water. Obtained deposit was filtered, washed out by distilled water and ethanol and dried in vacuum at 70 °C and 133 Pa.

Discussion

Obtained polyetherimides are of interest as easily processed polymers on the standard equipment. Synthesized polyetherimides can be used as thermoreactive polymers, their mixes with other heterochained and heterocyclic polymers also are of interest with the purpose of obtaining the composite materials with high chemical and fire resistance.

References

[1] Beck, J.R., Sobczak, R.L., Suhr, R.G., Yahner, J.A., *J. Org. Chem.*, 39:1839 (1974).

[2] Relles, H.M., Orlando, C.M., Heath, D.R. et al., *J. Polymer Sci. Polymer Chem.*, 15:2441 (Ed. 1977).

[3] Williams, F.J., Relles, H.M., Donahue, P.E., *J. Org. Chem.*, 42:3419 (1977).

[4] Williams, F.J., Relles, H.M., Donahue, P.E., Manello J.S., *Ibid.*, 42:3425 (1977).

[5] Williams, F.J., Donahue, P.E., *Ibid.*, 42:3414 (1977).

[6] Takekoshi, T., Wirth, J.G., Heath, D.R. et al., *J. Polymer Sci. Polymer Chem.*, 18:3069 (Ed. 1980).

[7] Planka, Z.J., Albreht, V.M., *Visokomol. Soed.*, 7A:2117 (1965).

[8] Berlin, A.A., Liogonkij, B.I., Shamraev, G.M., Belova, G.V., *Izv. AN USSR. Ser. Him.*, 5 (1966).

[9] Korshak, V.V., Rusanov, A.L., Pavlova, S.A., etc., *Dokl. AN.* USSR, 221:1334 (1975).

[10] Brown, B.R., Todd, A.R., *J. Chem. Soc.*, 4:1280–1285 (1954).

In: Polymers, Polymer Blends, Polymer Composites... ISBN 1-60021-168-2
Eds: A.K. Mikitaev, M.K. Ligidov et al., pp.93-97

Chapter 11

PROPERTIES OF THE FILLED ACRYLIC POLYMERS

O. A. Legonkova*
Moscow State University of Applied Biotechnology

ABSTRACT

In the present work physical and mechanical properties of inflated with organic and inorganic fillers acrylic polymers in wide range of ratio of components in inert (air) and aggressive (water) mediums are discussed. Electronic micrscopy and infra red spectroscopy was used for explanation of the received data.

Keywords: fillers, compounded mix, physical and mechanical properties.

Acrylic polymers are polymers that are received through interaction of polyatomic alcohols, phenols and acrylic acid. The presence of one or more isolated binary bonds stipulates for their high reactionary ability. Properties of the received polymers depend on the nature of substances, entering into chemical reactions. To get materials with wide range of properties one can modify components, for example, via mixing acrylic acid with unsaturated aliphatic or aromatic bibasic carbon acids. As a result, we get acrylic polymers of different forms and structure.

Traditionally acrylic polymers are used in the form of dyes, lacquers, enamels [1,2]. But properties of the filled acrylic polymers are studied insufficiently, though filled with organic or inorganic fillers polymer compositions that have various chemical nature and structure, are being widely employed [3].

In this work polymers' films made out of acrylic dispersions, such as "Forcit", Finland, A10 (acrylic-styrene dispersion), A2001, A30, Lentex A4, Russia, were investigated. Acrylic dispersions were mixed with organic and inorganic fillers. A well-known in Russia fertilizer "Rastvorin A" (which is a mixture of salts, i.e. sulphate, phosphate and nitrate of ammonium and magnesium) was taken as inorganic filler, and waste of grain production (which is

* Moscow State University of Applied Biotechnology, mailto: ol@te-ka.ru

organic powder with particles dimension 64-240 microns, poured density 350 kg/cm^3) was taken as organic filler.

In the table 1 physical and mechanical characteristics of the inflated with organic and inorganic fillers acrylic compounded mix based on A4 acrylic dispersion are presented. The same behavior of samples based on the other acrylic dispersions was revealed.

Table 1

Content of A4 acrylic dispersion, %	Content of organic filler, %	Content of inorganic filler, %	Strength, MPa	Deformation at break, %
100	0	0	1,5	1000
90	0	10	1,3	1200
80	0	20	1,0	1400
60	0	40	0,8	800
90	10	0	1,6	900
80	20	0	1,7	400
60	40	0	2,5	150
40	40	20	0,5	2000

As it's shown in the table 1, the increase of organic filler content in compounded mix leads to deterioration of flexibility. The introduce of organic filler brings the strength increase. While the increase of content of inorganic filler (till 20%) brings noticeable deformation, though the durability decreases.

On the 1-3(th) figures electronic microscopical paintings of the splitted surface of the samples: the initial polymer matrix (fig.1); polymer matrix inflated with inorganic filler (fig.2); polymer matrix inflated with organic filler (fig.3), are presented.

Fig. 1. Electrical microscopical picture of the splitted surface of initial acrylic polymer matrix based on acrylic dispersion A4

As it's shown on the fig.1-3, compounded mix inflated with organic filler has more homogeneous structure then mix with inorganic filler which has substantively dissimilar structure. This dissimilar structure includes crystallites of salts, and with increasing of filler content this homogeneousness became more vivid and explain the decrease of durability of the compounded mixes, table 1.

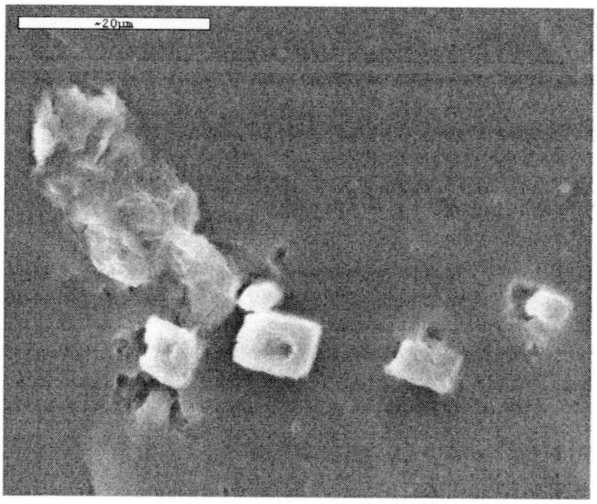

Fig. 2. Electrical microscopical picture of the splitted surface of the inflated with inorganic filler (20%) acrylic polymer matrix based on acrylic dispersion A4

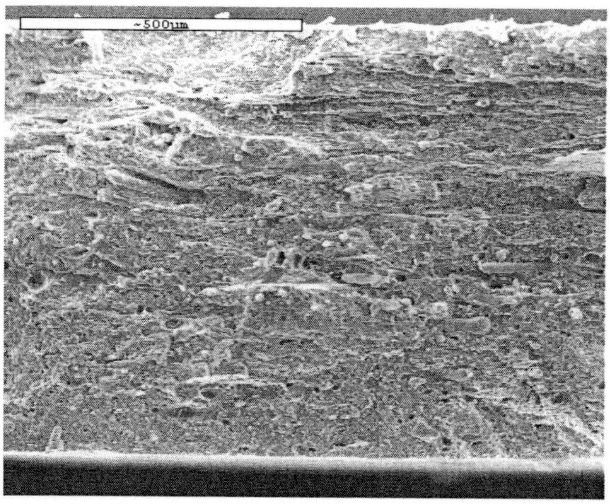

Fig. 3. Electrical microscopical picture of the splitted surface of the inflated with organic filler (20%) acrylic polymer matrix based on acrylic dispersion A4

According to infra-red spectroscopy, inorganic filler is inert regarding to the polymer matrix as spectrums of absorption of not filled polymer films and samples inflated with inorganic filler are the same.

While having a small content of organic filler in polymer matrix and as a consequence, absence of aggregation of powdery particles, fig.3, braking of demolition processes may take place [4], which leads to the durability increase of heterogeneous system, table 1.

The next step was the investigation of influence of water on changing of physical and mechanical properties of compounded mixes, tables 2 and 3.

Table 2. Durability changes (MPa) in polymer matrixes based on different acrylic polymers after exposition in water

Grades of acrylic polymers	Exposition of samples in water, days				
	0	5	10	14	18
A10	1,5	1,4	1,3	1,2	0,5
A2001	4,2	3,8	2,0	1,8	1,2
A30	11,0	3,5	3,0	1,7	1,1

Table 3. Durability changes (Mpa) in filled systems

Content of acrylic polymer A4, %	Content of organic filler, %	Content of inorganic filler, %	Exposition of samples in water, days				
			0	5	10	14	18
90	0	10	1,3	3,0	3,5	3,6	3,8
80	0	20	1,0	2,3	2,1	2,3	2,2
60	0	40	0,8	1,1	0,9	0,8	0,9
90	10	0	1,6	1,7	3,0	3,5	4,0
80	20	0	1,7	1,7	2,5	2,6	2,5
60	40	0	2,5	1,6	0,8	0,5	0,4
40	40	20	0,5	0,9	1,0	0,5	0,2

As it's shown, the durability of all polymer mixes based on different acrylic dispersions decreases with time in water. Though the deformation at break increases in 3-5 times. So, we can say about plasticizing influence of water for these polymers.

As you can see, table 3, the increase of organic filler content up to 40% brings the durability increase. Contrary to the behavior of organic filler, the inflation with inorganic filler leads to the decrease of durability of samples.

At small content of both fillers (till 20%) the strength of the samples increases with exposition in water. The increase of durability during exposition of samples in water can be explained with revealed changes in relaxation properties of the investigated systems, displayed flexibility and manifested ability to realization of orientation in the process of deformation [5].

Thus, fillers insert their own amendments in mechanical properties of the compounded mixes independently from each other, i.e.: the presence of organic filler brings an increase of durability at small exposition in water, while inorganic filler improves flexibility of the compounded mix.

References

[1] *Receipt, properties and application of water soluble acrylic copolymers*, Moscow, "Chemistry", 1982, 280p.

[2] Glebov, I.P. *Use of production waste from acrylic polymers as fertilizers*. Leningrad, 1983, 50 p.

[3] Lipatov, J.S. *Physical and chemical bases for filling of polymers*. Moscow, "Chemistry", 1991, 260 p.
[4] Erkova, L.N., Chechik, O.S. *Latexes*, Leningrad, "Chemistry", 1983. 223p.
[5] Lipatov, J.S. *Colloid chemistry of polymers*, Kiev, Naukova dumka, 1984, 340p.
[6] Askadskii, A.A. *Physical chemistry of polyarylates*, Moscow, "Chemistry", 1968, 210p.

In: Polymers, Polymer Blends, Polymer Composites... ISBN 1-60021-168-2
Eds: A.K. Mikitaev, M.K. Ligidov et al., pp.99-105

Chapter 12

POLYSULFONETHERKETONES ON THE OLIGOETHER BASE, THEIR THERMO- AND CHEMICAL RESISTANCE

Zinaida S. Khasbulatova[1*], Luiza A. Asuyeva[2], Madina A. Nasurova[2], Arsen M. Kharayev[2] and Gennady B. Shustov[2]
[1]Chechen State University, Grozny
[2]Kabardino- Balkarian State University, Nalchik, 173 Chernishevskogo Street

ABSTRACT

In the article are described package - sopolysulfonetherketones, manufactured on the oligoether base and are given the research results of their thermo- and chemical resistance study in agressive medium (acids and alkali). Oligomers are manufactured by high temperature polycondensation method in aprotonic dipolar solvent medium - dimethylsulfoxide (DMSO) - within inert gas atmosphere (nitrogen). The structure of produced oligomers is proved by the results of element analysis and IK - spectroscopy. It is demonstrated that synthesized polysulfonetherketones has a good thermoresistance and resistance to acid and alkali reaction.

Keywords: oligosulphone (OS), oligoketon (OK), oligosulphoneketon (OSK), polycodensation, package - sopolysulfonetherketones, dian (D), phenolphtalein (PH), compound ether links, ordinary ether links, thermoresistance, chemical resistance.

In different fields of technics widely used such polymer materials as: polyarilates, polysulfones, polyketons. Each one of them is characterised by definite advantages over the others. As a fact, they have their own drawbacks.

To combine positive properties of different classes of polimers in one material lately widely started to use oligomers, that contain in their chain links of this class of polimers.

---

* Correspondence to: Zinaida S. Khasbulatova, Chechen State University, Grozny, 33 Kievskaya street, h_Zina @ email. ru

So, with the objective to improve some properties of polysulfones and polyetherketones are manufactured and examined some properties of number of package sopolymers [1-7, 8, 9], which attract a wide interest from different fields of industry as a thermoresistant materials of constructive and electroisolative application. With the objective to manufacture package-sopolysulfonetherketones with high molecular mass, thermoresistance, and good physico-mechanical characteristics are synthesized oligoketons (OK), oligosulphoneketons (OSK), oligosulphones (OS) of different structure and condensation degree.

Oligosulphoneketon, oligosulphone, oligoketon synthesis was conducted by means of high temperature polycondensation in aprotonic dipolar solvent medium - dimethylsulfoxide (DMSO) - within inert gas atmosphere (nitrogen).

Test reaction was conducted for diphenylolpropane (DPHP) disodium salt and $4,4^1$ dichlorodiphenylsulfone (DCHDPHS) in OS-D case, and phenylphtalein disodium salt (PHPH) and $4,4^1$ - dichlordiphenylsulfone in OS-PH case.

In case of getting oligoketons instead of $4,4^1$ - dichlorodiphenylsulfone were used the corresponding molar quantity of $4,4^1$- dichlorbenzophenone (DCHDBPH).

In case of getting oligosulfoneketones (OSK) test reaction was conducted for diphenylolpropane disodium salt (DPHP) (or phenolphtalein) and for a mixture 50:50 molar % of dichlordiphenylsulfone and dichlor diphenylbenzophenone.

Oligosulphone synthesis is conducted in molar proportion DPHP: DCHDPHS-2:1 (OS-1D); 6:5 (OS-5D); 8:7 (OS-7D); 11:10 (OS-10D); 21:20 (OS-20D) and PHPH:DCHDPHS-2:1 (OS-1PH); 11:10 (OS-10D); 21:20 (OS-20PH).

Some oligosulphone, oligoketon, oligosulphoneketon properties are demonstrated in Table 1.

Table 1. Oligosulfone, oligosulfoneketone, oligoketone properties

Oligomers	Polycondensation degree	n_{pr} m^3 / kg	Output, %	Melting t°, K	Estimated molar mass	Hydroxile group contents,%
OS-1D	1	0,002	98	358-361	670,84	5,07 5,08
OS-20D	20	0,024	98	458-462	9078,05	0,38 0,32
OS-1PH	1	0,002	98	475-778	850,90	4,00 4,03
OS-20PH	20	0,025	99	565-573	10969,88	0,31 0,29
OK-1D	1	0,003	98	402-403	634,78	5,35 5,40
OK-20D	20	0,020	99	440-449	8358,01	0,40 0,45
OK-1PH	1	0,003	98	462-473	814,85	4,17 4,20
OK-20PH	20	0,022	99	528-533	10248,77	0,33 0,30
OSK-1D	1	0,006	98	410-416	1077,30	3,15 3,20
OSK-20D	20	0,022	99	452-467	17208,58	0,20 0,18
OSK-1PH	1	0,004	98	471-473	1347,43	2,52 2,62
OSK-20PH	20	0,013	99	536-544	20900,39	0,16 0,22

The general diagram of oligosulphone, oligosulphoneketon, oligoketon synthesis you can draw as follows:

${}_{n+1}$ NaO-R-ONa+ ${}_{n}$ Cl-R'-Cl —> NaO-R-(O-R'-O-R) - ${}_{n}$ONa, where R=

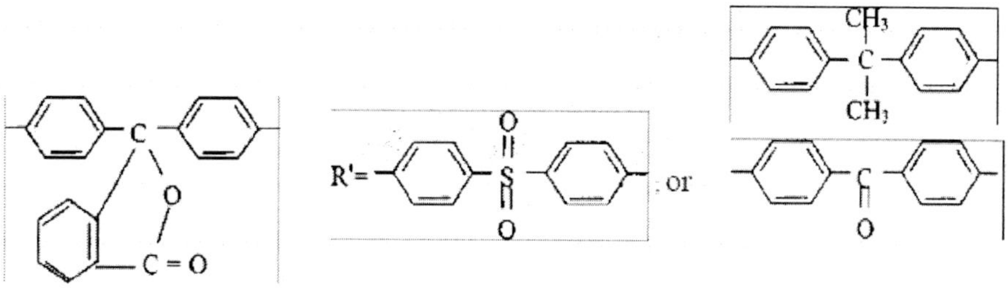

their equimolar compound (50:50)

The structure of manufactured oligomers is proved by the results of element analysis and IK- spectoskopy. The presence of absorbtion stripes in IK- spectres, corresponding to ordinary ether links in fields 980,1014,1045 cm.$^{-1}$ to isopropylidene group in dian remainder 1290, 1365, 1385, 1415-1456; 1930-1980cm.$^{-1}$ (in case of dian oligomers) to lactone group 1750-1780cm. $^{-1}$ (in case of phenolphtalein oligomers), to hydroxile groups 3650-3740cm.$^{-1}$ to sulphonylic group 1150-1170, 1215,1245-1285, 1820cm.$^{-1}$ (for oligosulfones) and ketogroup 1610-1650cm.$^{-1}$ (for oligoketons and oligosulfoneketons) testifies the formation of oligosulfones, oligosulfoneketones and oligoketones.

Since from 1960-s aromatic ethers attract attention of different researchers for their thermoresistance, also for a number of valuable properties [10-12].

It is worthy to note, that according to their thermo- and chemical resistance aromatic ethers outdo polyarilates, but the latter is superior in heat resistance.

Study of thermoresistance of the condensational type of polimers is a matter of great importance nowadays. In this connection great interest attracted the research of the thermoresistance of polysulfonetherketones (PSEK), synthesized by us. The results of thermogravimetric PSEK analysis are given in Table 2.

Table 2. Thermic properties of polysulfonetherketones

№ п/п	Primary compounds		Thermoresistance, K		
	dioxy-compounds	dichloranhydrids	2%	10%	50%
1	OSK-1D	DCHATK/DCHAIK	688	760	970
2	OSK-10 D	DCHATK/DCHAIK	699	770	988
3	OSK-20 D	DCHATK/DCHAIK	720	777	993
4	OK-1 D+OC-1 D	DCHATK/DCHAIK	683	753	833
5	OK-10 D+OC-10 D	DCHATK/DCHAIK	693	763	870
6	OK-20 D+OC-20 D	DCHATK/DCHAIK	710	765	873
7	OSK-1PH	DCHATK/DCHAIK	692	770	978
8	OSK-10PH	DCHATK/DCHAIK	690	778	986
9	OSK-20PH	DCHATK/DCHAIK	704	782	998
10	OK-1PH+OC-1PH	DCHATK/DCHAIK	696	766	970
11	OK-1PH+OC-10PH	DCHATK/DCHAIK	690	770	978
12	OK-20PH+OC-20PH	DCHATK/DCHAIK	698	776	995

Analysis data, given in Table 1, demonstrate that polysulfonetherketones are characterised by high thermic indicators and outdo polyarylates, polyarylatesulfones, polyetherketones. It's worthy to note, that 2% mass loss for all synthesized PSEK keeps on the 673 level and higher.

Among polysulfonetherketones on the dian oligosulfoneketon base (OSK-D), equimolar mixture of iso-and terephthalic acid dichlorinhydrins the most inferior thermoresistance has BSP on the OSK-1 D base. According to the lengh grow the increase of PSEK thermoresistance is observed, which explained by the reason that with the OSK lengh growth the fewer unstable compound ether links are left. The same order is observed among the number of oligosulfoneketones (OSK) on the phenolphtalein base. In these polysulfonetherketones the saturation of polimer chain by thermoresistant ordinary ether links is observed and a portion of unstable compound ether links, as it was noted before, falls down abruptly, moreover, with the increase of primary OSK data package density in polysulfonetherketones is increases. Probably, these three factors promote the predictable growth PSEK thermoresistance inthe line. Also, has been studied thermic characteristics in polysulfonetherketone line on the equimolar compound dian oligosulfone base (OS-D) with oligoketones (OK-D) and on the phenolphtalein oligosulfone compound base (OS-PH) with oligoketones (OK-PH). The comparison of these two rows shows that polimers on the equimolar compound base of phenolphtalein oligosulfones with the oligoketons demonstrate more high indicators of thermic resistance. Package sopolysulfonetherketoneon the OS - 20 PH+OK - 20PH base possesses the most thermoresistance t^0 of 2% and 10% mass loss of this BSP is accordingly equated to 638 and 776 K (Table 2).

With the length growth of primary oligomers in every group of polimer polysulfonetherketones is observed a considerable increase of polimer thermoresistance, the reason for this is, evidently, that with the lenth growth of OSK, OS and OK in polysulfonetherketones reduces the quantity of unstable compound links that positively influences thermic resistance.

Most of the thermoresistant polimers are distinguished by their considerable chemical resistance in acids and alkalis [13]. A great interest in study of package sopolysulfonetherketone chemical resistance attributes to this fact.

Polyethers are stable in mineral and organic acids, except for concentrated sulphur acid, in diluted alkalis and some oxidizers [14]. Polimer chemical resistance can be increased by diminishing the concentration of accessable chemically unstable links, by means of injecting the macromolecules with substitutes inhibitory for aggressive medium components to have access to chemically unstable links.

Tests for film PSEK specimens have been conducted in 10%, 30% concentrated sulphur acid, 10% and concentrated saline acid (36,5%), in 10%,50% sodium hydroxide solution. The research results are given in Table 3. The Table demonstrates that polysulfonetherketones on dian OSK base display good stability in diluted sulphur acid solutions, as well as in concentrated saline acid. In concentrated alkali PSEK is exposed to destruction, that is probably, is due to the presence in the polysulfonetherketones chain chemically unstable compound ether links. This justifies the fact, that PSEK based on short OSK and saturated by these links they are sooner distructed than PSEK on OSK-10 and OSK-20. More than that, this speed of polimers is caused by low package density of PSEK on short oligosulfoneketone base.

Table 3. Polysulfonetherketones mass change dependance from exposure duration in agressive medium

№ п/п	Primary oligomers	Time of exposure, sec-10^5	Mass change, %				
			H_2SO_4		HCl	NaOH	
			10%	30%	36,5%	10%	50%
1	OSK-1 D	0,864	0,28	0,18	0,59	0,71	0,26
		1,728	0,56	0,72	1,42	2,38	-0,70
		3,456	0,63	0,91	2,01	2,42	-1,83
		13,824	0,79	1,01	2,32	0,11	-9,01
2		0,864	0,20	0,15	0,42	0,61	0,17
		1,728	0,34	0,53	1,21	1,82	0,62
		3,456	0,68	1,19	1,89	2,17	-1,52
		13,824	0,72	1,26	1,95	0,97	-7,11
3	OSK-20 D	0,864	0,14	0,13	0,23	0,43	0,10
		1,728	0,28	0,32	0,92	1,31	0,47
		3,456	0,48	0,83	1,91	1,62	-0,83
		13,824	0,59	0,97	1,96	1,47	-5,21
4	OSK-1 PH	0,864	0,36	0,81	0,99	0,40	0,41
		1,728	0,59	0,80	1,57	0,56	-0,59
		3,456	0,75	1,66	2,06	0,90	-0,87
		13,824	0,95	1,83	2,19	1,02	-9,75
5	OSK-10PH	0,864	0,38	0,55	0,66	0,47	0,67
		1,728	1,00	0,91	1,31	2,00	-0,31
		3,456	1,91	1,60	2,40	0,46	-0,06
		13,824	2,00	1,66	2,48	0,82	-7,93
6	OSK-20 PH	0,864	0,18	0,40	0,63	0,55	0,21
		1,728	0,85	0,97	1,74	1,60	-0,42
		3,456	1,45	1,29	1,89	1,29	-0,12
		13,824	1,48	1,39	2,15	1,63	-6,40
7	OK-1 D+OS-1 D (50:50)	0,864	0,27	0,30	0,40	0,31	0,17
		1,728	0,64	0,78	0,88	0,62	-0,16
		3,456	0,98	0,43	1,34	1,02	-0,63
		13,824	1,13	0,14	1,45	1,14	-7,77
8	OK-1 0 D+OS-1 OD (50:50)	0,864	0,28	0,17	0,24	0,21	0,20
		1,728	0,54	0,40	0,56	0,50	0,25
		3,456	0,83	0,66	0,74	0,59	0,16
		13,824	0,96	0,65	0,75	0,60	-1,12
9	OK-20 D+OS-20 D (50:50)	0,864	0,28	0,17	0,24	0,21	0,20
		1,728	0,54	0,40	0,56	0,50	0,25
		3,456	0,83	0,66	0,74	0,59	0,16
		13,824	0,96	0,65	0,75	0,60	-1,12
10	OK- 1PH+OC-1PH	0,864	0,62	0,77	0,70	0,48	0,57
		1,728	0,89	0,92	1,69	0,72	-0,72
		3,456	1,45	1,66	2,40	1,02	-0,91
		13,824	2,20	2,10	3,00	-1,22	-11,50
11	OK-10PH+OS-10PH	0,864	0,20	0,47	0,69	0,59	0,30
		1,728	0,89	0,96	1,62	1,65	-0,51
		3,456	1,30	1,36	1,90	1,32	-0,20
		13,824	1,55	1,47	2,00	-0,01	-7,40

The difference of PSEK package density is underlined clearly in this polimer group when they swell in acid solution. Diluted sulphur acid solutions don't influence PSEK, their swelling is generally due to water absorption. In 10% sulphur acid mass stabilization follows after swelling, that demonstrates the absence of solvability, all the more so of destruction. Preservation of resistant properties and permanence of indicated viscosity of BSP also justifies the absence of distructive processes.

Manufactured polysulfonetherketones mostly swell in concentrated saline acid solution. This is due to the fact, that saline acid belongs to electrolytes with high pressured steem which penetrate into polimers that have coefficient close to water coefficient. The swelled PSEK specimens even after $13{,}82x10^5$ seconds keeping in concentrated saline acid solution don't loose mass. (Table 3).

The comparison of two PSEK rows on the oligosulfoneketone base demonstrated that polysulfonetherketones on the phenolphtalein OSK base have greater swelling capacity, it, probably, depends on the reason that package density BSP on phenolphtalein OSK base is less in comparison with polimers on dian OSK base.

Combination of dian oligoketons with dian oligosulfones improves the PSEK stability in concentrated and diluted alkali solution, that is, evidently, connected with the presence in their structure chemically stable ordinary ether links. They are also stable in diluted sulphur and concentrated saline acids.

Chemical stability is increased with the primary oligoketone and oligosulphone lenth growth. BSP on the OK-1D + OS-1D base is less chemically stable as it outstands by its greater content of compound ether links.

Phenolphthalein BSP on the oligoketon and oligosulfone compound base are exposed to destruction in 10% alkali, have greater swell per cent in diluted sulphur acid solution and in concentrated saline acid solution in comparison with PSEK on the dian oligomers compound base.

The most chemically stable in polimer group is PSEK on OK-20-PH+OS-20PH, probably, it's due to high concentration of ordinary ether links in its structure and reduced concentration of unstable compound ether links.

Accordingly, synthesized polysulfonetherketones are characterised by high thermoresistance qualities, demonstrate chemical stability and can be used as polimer materials of constructional and film application.

REFERENCES

[1] Huml, J. Doupovkova, J. *Polysulfone nogu drei guntetickych pruskeric - plast*, hmoty a kanc.,1970, 7, №4, 102-106.

[2] Morneau, Ct. A thermoplastic polyanylsulphone that can be used at 500f. *Mod. plast.*, 1970, 47 №1, 150-152, 157.

[3] Storojuk, I.P. Valetzky, P.M. Formation order and properties of poryarylensulfonoxides. — Science and technics achievements. *Chemistry and technology of high molecular compouds*. B.2 1978, p. 127-176.

[4] Bensen, B. A. Brinder, R. P., Vogel, H. A. *Polymer 360,* A. Termoplastic Por Use at 500 F, presented at SFE Antes, Detroit, Michigan, May 17, 1967

[5] Jolem, A. Phenelene Termoplastic for use at 500F *SPE Journal,* 23,33, July, 1967

[6] Besset, H. D. Fazzari, A. M., Staub, R. B. - *Plast Technol.*, 11 (9), 50,1965

[7] Jaskot, E. S. -*SFE Journal*, 22, 53, (May, 1966).

[8] Criordi, E. 0. Termoplastico de endenharis ideal para as condicoes Erasu Leiras. - *Rev. guin. ind.*, 1971. 40. N° 470,16-18.

[9] Korshak, V.V. Storojuk, I.P. Mikitayev, A.K. Polysulfones — sulfonecontaining polimers. — Collection. *Polycondensational processes and polimers.* Nalchik, 1976, p.40-78.

[10] White, D. M. In Comprehensive Polymer Science. *The Synthesis, Characterization, Reactions, and Applications of Polymers.* V. 5. Pergamon Press, Oxford, 1989.P. 473.

[11] Staniland, P. A. In Comprehensive Polymer Science. *The Synthesis, Characterization, Reactions, and Applications of Polymers.* V. 5. Pergamon Press, Oxford, 1989.P. 483.

[12] Parodi, F. In Comprehensive Polymer Science. *The Synthesis, Characterization, Reactions, and Applicatipns of Polymers.* V. 5. Pergamon Press, Oxford, 1989.P. 561.

[13] Vorobyova, G.Y. Chemical stability of polimer materials, *M.:Chemistry*, 1981.-133 p.

[14] Vorobyova, G.Y. Corrosion resistant materials in agressive medium of chemical manufacture. *M.: Chemistry*, p.1975-173.

In: Polymers, Polymer Blends, Polymer Composites... ISBN 1-60021-168-2
Eds: A.K. Mikitaev, M.K. Ligidov et al., pp.107-113

Chapter 13

The Mechanism of Inhibition Thermooxidation Destruction of PBT by Polymer Azomethines

B. S. Mashukova[1]*, T. A. Borukaev[1], N. I. Mashukov[1] and M. A. Mikitaev[2]
[1]Kabardino-Balkarian State University, Nalchic City, Russia
[2]State Scientific Institution"Compound centre", Moscow City, Russia

Abstract

All possible mechanisms of inhibition of thermooxidation destruction of polybutilenterephthalate by polyazomethines on basis of triarylmethane line diamines have been examined. The dependence of stabilizing properties of polyazomethines from their chemical structure and molecular mass has been shown.

Keywords: polybutilenterephthalate, stabilization, mechanism, polymer azomethines.

It has been mentioned before that [1] there are potential reactivity centrals at synthesized polyazomethines (PAM). Each of them can take part in radical processes, which are a part of summary thermooxidation destruction of: polybutilenterephthalate (PBT), which acts mainly according to chain-radical mechanism.

It is important contribution of azomethines groups of these PAM into inhibition of thermooxidation destruction of PBT deserves attention. Particularly inhibiting ability of such compounds is stipulated by conjugate effect –CH=N- bonds [2]. The mechanism of action –C=N- bonds consists of the following: π-electrons of these systems are clever to turn into higher energetic level when activated, which can effectively accept free radicals. The acceptation of free radicals grow especially at high temperatures with help of –CH=N-bonds.

*Kabardino-Balkarian State university, 360004,Russia, Nalchic city, Chernyshevsky st. 173; boruk-chemical@mail.ru, Mahsuk_bela@mail.ru

Next reactivity central of PAM, which is clever to inhibit radicals processes is end amino groups:

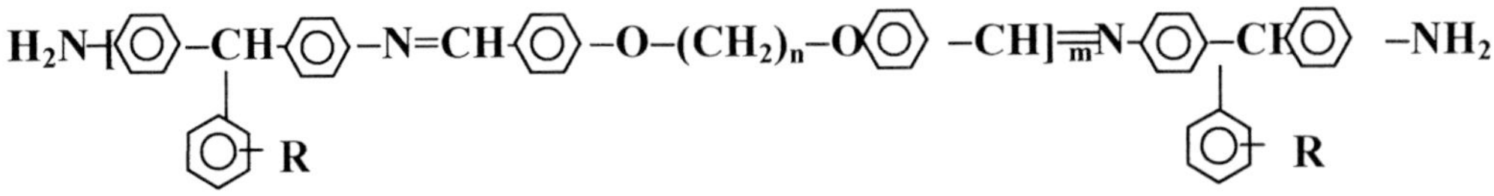

where, R- H; CH_3; NO_2. n=4,6,8,10.

The mechanism of their action is as follows. Firstly, it is possible formation of complex with transfer of charge between aromatic nucleus of inhibitor and peroxide radical, which is responsible for development of oxidative processes [3]. In consequence of destruction of this complex molecular products are formed. Secondly, another mechanism, which is more investigated is a process of inhibition, which accompanied by isolation of hydrogen from aminogroup on scheme[3,4]:

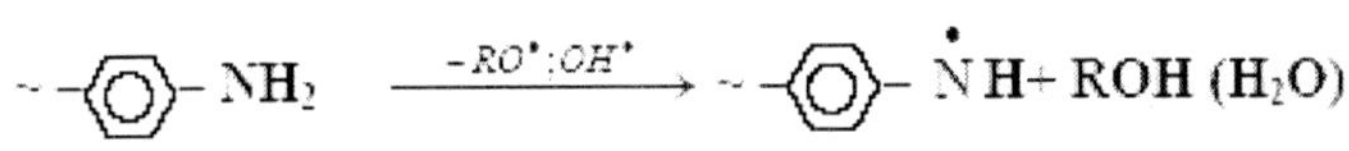

During this process a new radical stabilizes thanks to its conjugation with aromatic nucleus.

Here it is important to note also, that availability of two end aminogroups with macromolecule of PAM allows to deactivate radicals in the same degree on certain mechanism, which are formed with disintegration of hydro peroxides:

$$ROOH \xrightarrow{K} RO^{\bullet} + OH^{\bullet}$$

In its turn formed radicals ($RO^{\bullet}$and $OH^{\bullet}$) are deactivated by end aminogroups of PAM macromolecular according to scheme shown above.

And, finely, the third reactivity central in the molecular of PAM is a labile atom of hydrogen of triarylmethane grouping. Availability of this atom allows us to assume about possible inhibition of radical processes because of isolation this hydrogen atom according to the following:

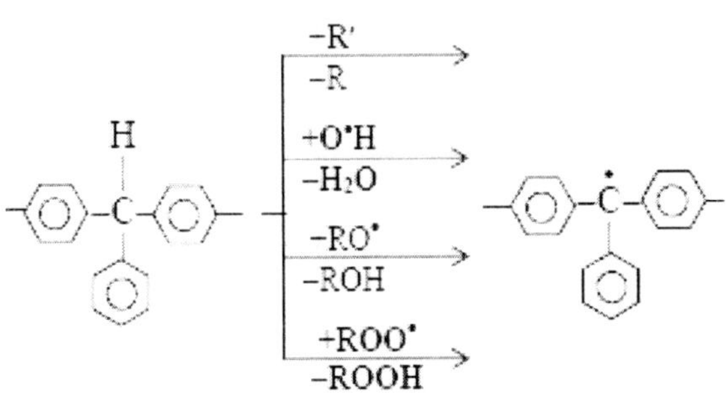

The high reactivity of hydrogen atom of triarylmethane fragment has been proved with its act for brom atom. For this purpose bromsuccinamide, which is used as a rule, for act of mobile hydrogen atoms ore for joining to double bond was used in this work. Reaction takes place according to the following scheme:

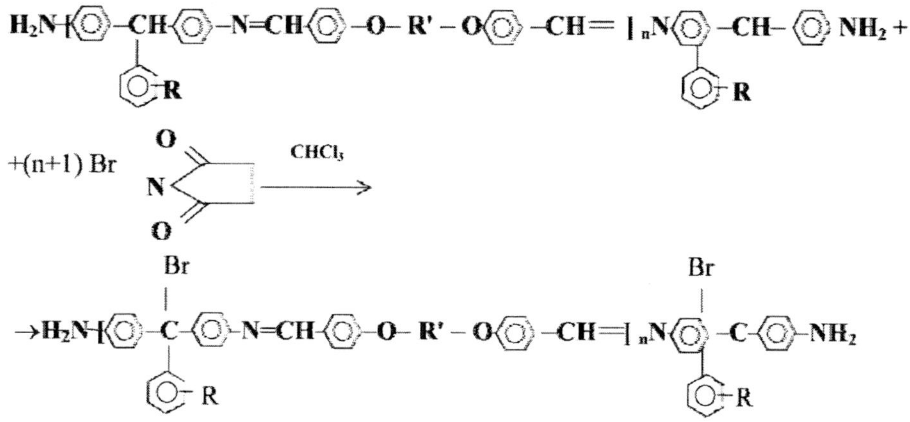

where R=H, CH_3; R'=$(CH_2)_m$; m=4, 6, 8.

The formation of such compound (bromcontent PAM) was acknowledged with help ^{1}H-NMR(nucleated-magnetic resonance), infrared spectroscopy and element analysis.

It received triarylmethine radical is sufficiently stable[5], that is acknowledged with the help of EPR(electron-para-magnetic) (fig.1) Further life of these radicals depends on different condition. So, in consequence of high-temperature oxidation in the process of diffusion of oxygen in polymeric female triarylmethine radicals can join in reaction with it. The scheme of reaction interaction of triarylmethine radicals with oxygen can be shown like this[6]:

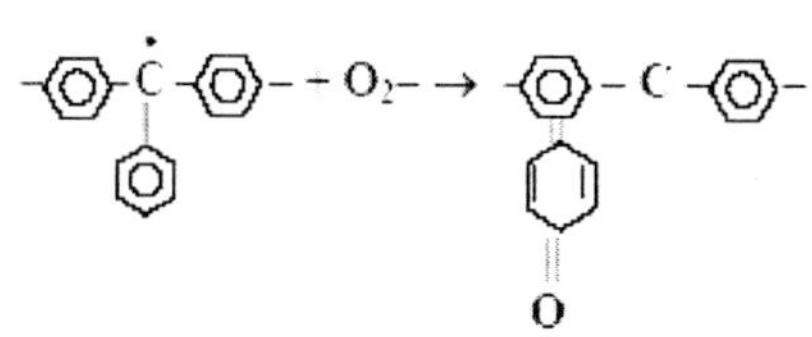

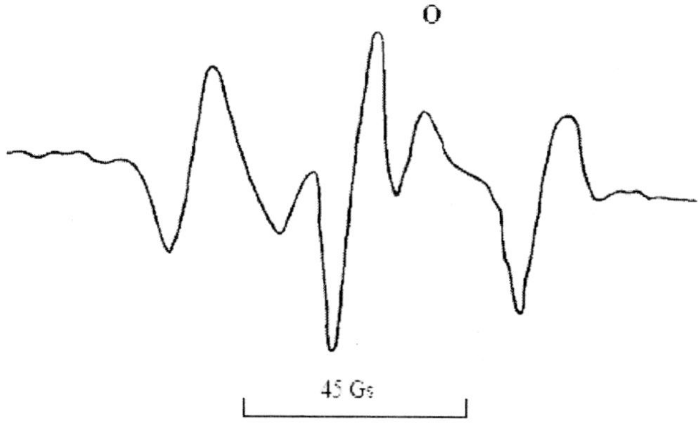

Fig. 1. Spectrums of EPR of PAM with triphrnylmethane radicals in basic chain. (Gs- gauss-magnetic induction)

Consequently, PAM with triarylmethine fragments in basic chain can inhibit thermooxidation destruction almost on all stage of oxidation of PBT or on account of interaction with radicals, giving molecular or stable products.

According to this the scheme of process of inhibited thermooxidation destruction can be shown in the following form:

I. Inhibition on stage of chains original:

$$RH + O_2 \xrightarrow{K_0} R^{\bullet} + HO_2^{\bullet}$$

$$RH + HO_2^{\bullet} \xrightarrow{K_0^{/}} R^{\bullet} + H_2O_2$$

$$R_3C^{\bullet} + O_2 \xrightarrow{K_{ПАМ}} \text{molecular products}$$

II. Inhibition on stage of chains development:

$$R^{\bullet} + O_2 \xrightarrow{K_1} ROO^{\bullet}$$

$$ROO^{\bullet} + R^{/}H \xrightarrow{K_2} ROOH + R^{/\bullet}$$

$$ROOH \xrightarrow{K_3} RO + \dot{O}H$$

$$ROO^{\bullet} \xrightarrow{K_4} \text{isomerization with burst of molecule}$$

$$R_3CH + R^{\bullet} \; or \; (O^{\bullet}H; RO^{\bullet}; ROO^{\bullet}) \xrightarrow{K_5} R_3C^{\bullet} + RH \text{ or } (H_2O; ROH; ROOH);$$

$$R' - NH_2 + RO^{\bullet} \; or \; (\dot{O}H) \xrightarrow{K_6} R\dot{N}H + ROH(H_2O) \qquad \text{Inert products}$$

III. Precipice of chain

$$R^{\bullet} + R^{/\bullet} \xrightarrow{K_7} R - R^{/};$$

$$R^{/}O_2^{\bullet} + R^{\bullet} \xrightarrow{K_8} R^{/}OOR;$$

$$RO_2^\bullet + R'O_2^\bullet \xrightarrow{K_9} \text{molecular products};$$

$$R_3C^\bullet + OH^\bullet \xrightarrow{K_{10}} R_3COH,$$

where $R_3C^\bullet$ = 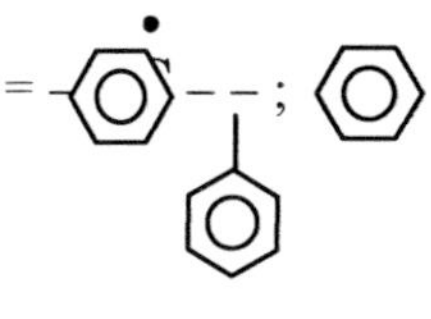

R_3CH = 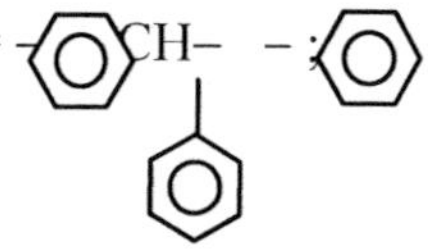

R′ – NH_2 CH = – –  –NH_2, –RṄH its radicals.

It should be noted that PAM used in work had different chemical structure. It was important to investigate influence of these structures on their stabilizing properties. So, using substituted diamines by synthesis of PAM, inhibiting activity of the previous ones lowers. In particular, PBT stabilized by PAM on basis 4,4$^{/}$ -diamino-4$^{//}$-p-methyltriphenylmethane and 4,4$^{/}$-diformyldi phenoxyhexane on significance of 2% mass loss is worse than industrial PAM on 10%.It is obviously that such conduct of this compound can be explained by big inclination to oxidation C-H-bond of methyl group. These sonfirmations indirectly confirm comparison of induction of period τ of PBT, stabilized PAM of different chemical structure. In particular, induction period τ can be valued as follows[7]:

$$\tau = f[U_0]/V_i$$

where f- ctoichiometric coefficient of inhibition, that is quantity of chains, which one molecule of inhibitor tears off. It successively joins reaction of isolation; $[U_0]$-concentration of inhibitor; V_i-speed of chains initiation.

With use of PAM with not substituted triarylmethane fragments in basic chain f=2(tear off hydrogen atom, accepted of oxygen).Then, we can substitute meaning f in the initial formula and induction period can be defined as follows:

$$\tau = 2[U_0]/V_i$$

For substitution of triarylmethane fragments in basic chain f will be equal to one, most probably, as reaction with oxygen are improbably. Then:

$$\tau` = [U_0]/V_i$$

From these equations we can define that PAM as polymer antioxidant on basic not substituted diaminotriphenylmethane are more effective. It can be shown as follows:

$$\tau/\tau` = (2[U_0]/V_i)/([U_0]/V_i) = 2$$

Consequently, PAM with not substituted triarylmethane fragments in basic chain surpass PAM on basic not substituted diamines two times on itself stabilizing properties.

Important significance have number of methylene groups in elementary link of macromolecule for stabilizing properties of PAM. Investigations of PBT thermofirmness, stabilized PAM with different number of methylene groups showed that with increase of "spacer" length is the result of temperature drop of 2% mass loss (fig. 2). And with change of thermofirmness of PBT+PAM compound are not with the transition from four methylene groups to six methylenes. Apparently, from standpoint of influence of the conformity facts (and as result, the change of agility and flexibility of macromolecule etc.), PAM is in more profitable position with six methylene groups then other. It causes higher stabilizing properties of this polymer.

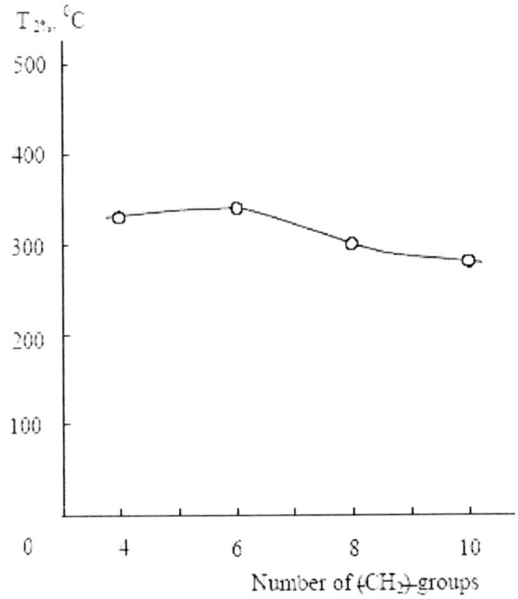

Fig. 2. The dependence T2% mass loss PBT+0,3% PAM compound from number of methylene groups in PAM

Besides, stabilizing properties of PAM are defined of molecular mass some extent. So, on the figure 3 the dependence of temperature of 2% mass lass of PBT+0,3% PAM compounds (PAM- on basic 4,4′ -diaminotriphenylmethane and 4,4′-diformyldiphenoxyhexane) from molecular mass of polymer antioxidant is shown.

From figure we can see that temperature significances of 2% mass loss of PBT=PAM compound is increasing visibly with growing of molecular mass of polymer antioxidant. With this temperature drop of 2% mass loss is happening not constantly, and to the definite meaning of molecular mass of antioxidant ($\overline{M}_w$=30000). It shows that stabilizing properties of PAM are constant with arrive at $\overline{M}_w$=30000 in future. Such conduct of stabilizing properties of PAM happens because with increase of molecular mass of PAM extension of conjugated chain does not happen but weakening of delocalization of π-electrons on chain takes place. This circumstance allows PAM to have sufficiently flexible and agility conjugated system, which together with aliphatic series creates more favorable conform conditions of interaction reaction proceed with free radicals.

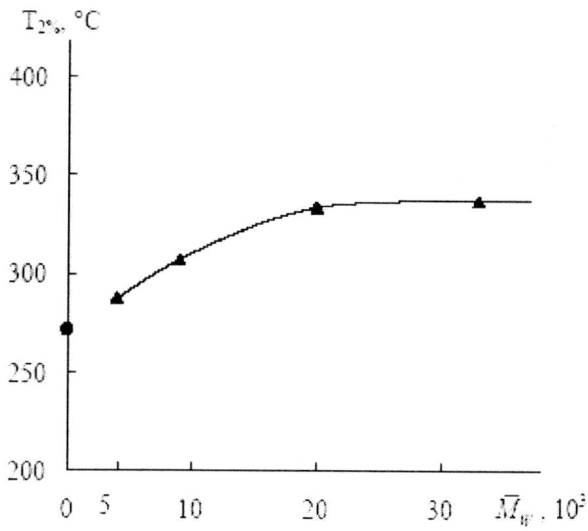

Fig. 3. The dependence T2% mass loss PBT+0,3% PAM compound from polyazomethines

Thus modifying chemical structure and molecular mass of PAM meaning we can elaborate polymer antioxidant, which can effectively inhibit thermooxidation of polymers and, in particularly, PBT.

References

[1] Borukaev, T.A., Tlenkopachev M.A., Mikitaev A.K. and others //*Plastic masses*. 1996. №2. P.17-19.

[2] Berlin, A.A.//*High-molecular weight compound*. 1971. V.13A. №2.P.276-293.

[3] *Aging and stabilization of polymers*. /by Kuzminski, A.S.-M.: Chemistry,1966.P.27.

[4] *Aging and stabilization of polymers*. / by Neiman,-M.: Science, 1964. 332 p.

[5] Braun, D., Faust J.// *Makromolec. Chem*. 1969. Bd. 121. №267. S.205-206.

[6] Rosantcev, E.G. Sholle, V.D. *Organic chemistry of free radicals*.-M.: Peace, 1979. P.99-128.

[7] Ankor, P. *Catalysis and inhibition of chemical reactions*. M.: Chemistry, 1966. 320 p.

In: Polymers, Polymer Blends, Polymer Composites... ISBN 1-60021-168-2
Eds: A.K. Mikitaev, M.K. Ligidov et al., pp.115-120

Chapter 14

AROMATIC BLOCK-CO-POLYETHERS AS PROSPECTIVE HEAT RESISTANT CONSTRUCTIVE MATERIALS

A. M. Kharayev, R. C. Bazheva and A. A. Chayka
Kabardino-Balkarian State University,
360004, Nalchik, Chernishevskaya St. 173, KBR, Russia

ABSTRACT

Aromatic polyetherketones of various structure and composition, which have good physical and chemical properties and good dissolubility in organic dissolvents were obtained by the method of acceptive and catalytic polycondensation. The reaction was carried out in two stages. In the first stage olygoketones of different degree of condensation were obtained on the basis of various bisphenols by the method of high temperature polycondensation. In the second stage block structure polyetherketones were synthesized with the use of the obtained olygoketones and dichloral hydrate iso- and terephtal acids by the accept-cathaletic polycondensation method. The connection between the olygoketone structure and polyetherketone properties was studied.

Key words: polyetherketone, olygoketone, heat resistance, polycondensation, viscidity, 1-2-dichlorineethane, triethylamin.

During last years chemist-synthesists and technologists were interested in aromatic polyetherketones (PEK) and polyetheretherketones (PEEK). The main feature in the composition of aromatic thermoplastic polyethers - polyetherketones and polyetheretherketones is the presence of one ether and one ketone group and two simple and one ketone group in their elementary bonds [1-3]:

$(\text{-Ar-O-Ar-C(O)-})_n$; $(\text{-Ar-O- Ar- O-Ar-C(O)-})_n$

The great interest to this class of polymers is explained by unique physico-mechanical, thermal, electrophysical characteristics of the given polymers conditioned by content phenyl groups in their main chains leading to a high degree of crystallite. According to these indices they exceed considerably many other thermoplasts [4-7]. Polyetherketones are partly crystal polymers their heat resistance depends on the temperature of glass transition (amorphity) and melting (crystallite) and it rises with decrease of molecular mobility. These polymers are insoluble in common dissolvent but they dissolve in concentrated sulphuric acid, they are hardly worn out, they preserve good electric properties in a broad interval of temperature and frequency, they are easily dyed with organic and inorganic pigments and may be glued together with various substratums in conditions of previous abrasive processing of the surface. PEEKs are easily processed by pressing, casting under preassure and extrousion. They are capable of another processing. The working temperature of these polymers with glass transition temperature (t_g) 415-420 K and melting temperature (t_m) 605-610 K is 570-580 K.

Taking into account the valuable properties of these polymers the probleme of perfection of the ways of synthesis and regulation the properties of these polymers becomes actual.

There is much in the literature [8-12] on the synthesis of this class of polyetherketones on the basis of 4,4′ – difluorinebenzophenon and well known bisphenols including dioxidephenilpropan and phenolphthalein. The duration of the synthesis (about 30 hours) and a high temperature (~550-570 K and higher) are serious shortcomings of these methods. Besides, even in these rigid conditions of synthesis other benzophenon halogenderivatives do not react actively. Besides, the obtained PEK and PEEK have bad dissolubilty in common organic dissolvents.

In this connection it was interesting to find the possibility of carrying out of the polycondensational process of polyether obtaining in much milder conditions with the use of 4,4′ – dichlorbenzophenon. As 4,4′- dichlorbenzophenon in dimetilsulphoxide medium at 423-433 K with 4,4′- dioxidephenylpropan and phenolphthalein does not give polymers with high molecular mass the obtaining of block structure polyetherketones was used.

With that end in view oligoketones of different composition and structure were obtained for their further use in polycondensational process. Olygoketones are synthesized according to the following scheme:

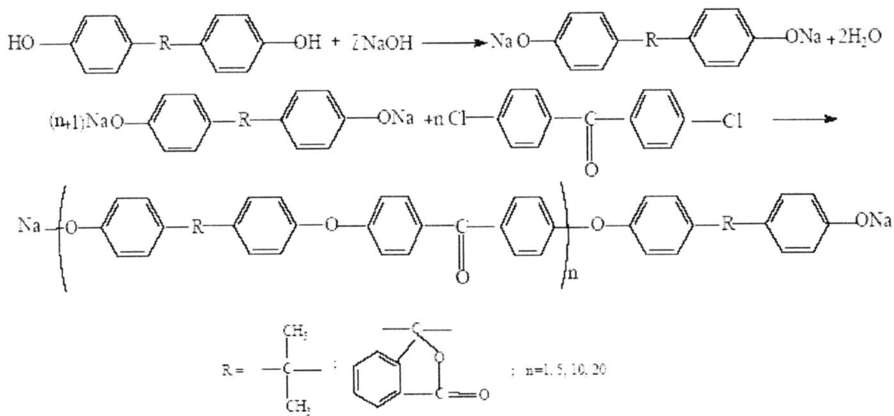

After precipitation in acidulated water and drying at 373 K olygoketones with ending hydrostrong groups were used for polymer obtaining. The structure of the obtained olygoketones is confirmed with elementary analysis and IK-spectroscopy data. The presence of absorption stripes, which correspond to simple ether bonds, isopropiden or lactone groups and also hydroxile and ketone groups witnwsses of plygoketone formation.

By acceptor-catalytic method of polycondensation in 1,2-dichlorethane medium block structure polyetherketones were obtained on the basis of obtained olygoketones and dichloranhydrides of iso- and terephtal acids mixture according to the scheme:

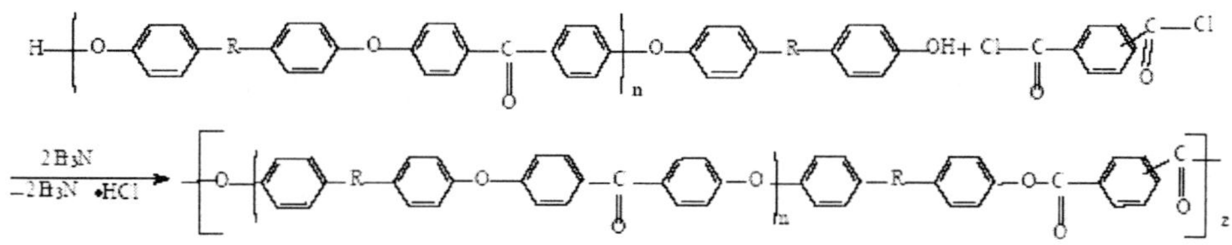

Optimal conditions of the synthesis of these polymers were defined: 1,2-dichlorrthane was a dissolvent, the temperature of the reaction was 295K, the time of carrying out was 60 minutes, consentration was 0,5m/l, thriethylamin was used as a catalist. PEEKs were obtained with a quantitative output and high indices of the given viscidity (n =1,2-2,5 d/g).

The composition and the structure of synthesized block structure PEEK are confirmed by the elementary analysis and IK-spectroscopy. The quantitative output and high value of the given PEEKs viscidity show indirectly the structure of the obtained polymers. Stripes of absorption are found on IK- spectra, which correspond to a simple ether bond (920-940sm $^{-1}$); to a complex ether bond (1000-1300 sm^{-1}) to diarilketone group (1600-1675 sm^{-4}) and stripes of absorption corresponding to hydroxide group were not found and this confirms the formation of PEEK of the expected structure.

The formation of PEEKs of the supposed structure is showed by the results of turbodimetric titration. The presence of PEEK solutions at one maximum on turbodimetrical titration curves confirms co-polymer formation but not homopolymer mixture (fig. 1).

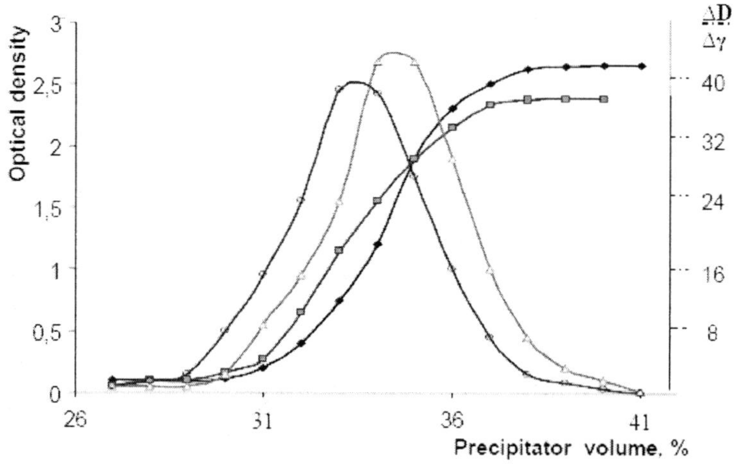

Fig. 1 The turbidimetric titration curves of PEEK obtained from OK-1D (♦, Δ) and PEEK obtained from OK-20D (■,○); integral curves (♦,■), differential curves (Δ,○)

Molecular masses of the synthesized polymers, which were measured with the ultracentrifuge by the method of approximation to equilibrium, are gradient from 40 to 140 thousand.

PEEKs are characterized by a good dissolubility in chlorinated organic dissolvent. The comparison of a series of PEEKs on the basis of dian and phenolphthalein olygoketones shows a better dissolubility of PEEKs on the basis of phenolphthalein olygoketones which is explained by their structure. PEEKs on the basis of phenolphthalein have a more friable structure and this increases dissolubility of polymers.

Some properties of synthesized PEEK are shown in Table 1.

Table 1. Properties of polyetherketones

PEEK obtained from:	Intrinsic viscosity (m^3/kg)	Glass transition temperature T_g, K	Melt temperature T_m, K	σ_p, MPa	ε_p %	Mass loss temp., K			Oxygen Index, %
						$T_{2\%}$	$T_{10\%}$	$T_{50\%}$	
OK-1D	0,24	423	513	75,0	15,0	670	733	863	3,22
OK-5D	0,19	425	525	77,2	14,1	678	740	855	3,00
OK-10D	0,16	430	545	80,1	12,5	682	765	873	3,00
OK-20D	0,13	433	563	85,4	8,2	693	813	886	2,97
OK-1F	0,25	448	568	80,1	19,9	678	780	663	3,43
OK-5F	0,18	453	578	82,4	18,0	680	780	863	3,40
OK-10F	0,14	468	590	84,4	16,0	682	785	863	3,25
OK-20F	0,12	473	603	90,0	10,3	690	790	860	3,06

where OK-1D, OK-5D, OK-10D, OK-20D are olygoketones on the basis of 4'4-dioxidephenilpropane; OK-1F, OK-10F, OK-20F are olygoketones on the basis of phenolphthalein with condensation degree n=15, 10 and 20 correspondently.

The investigation of PEEK properties showed that a noticeable increase of T_g and T_f with the increase of condensation degree of initial olygoketones is observed in a series of polymers. The comparison of thermo-mechanical characteristics of PEEKs on the basis of dian and phenolphthalein olygoketones shows that an introduction of volumetric card groups as a connecting group into PEEK structure, as it was expected, raises the amorphity and fluidity of the latter.

Low temperature of glass transition and fluidity of PEEK in comparison with PEK may be explained by the presence of a big amount of flexible simple ether bonds in the chain.

The comparison of durable properties of PEEKs shows that some increase of bursting durability with the increase of length of the initial OK is observed in some PEEKs. This may be explained by the density increase in the packing of the chain in PEEK on the basis of longer OK. All the PEEKs on the basis of phenolphthalein OK are characterized by higher results of durability properties in comparison with PEEKs on the basis of dian OK.

The obtained results on heat-resistance show that PEEKs are characterized by high indices of heat-resistance in the atmosphere.

The destructive process for all the models of polymers begins at 67OK and higher. The obtained series of PEEKs do not differ much, but a considerable increase of this characteristic is observed in some PEEKs on the basis of dian OK with the increase of the initial length of olygomers. This phenomenon can be explained as follows. On one hand in this line of PEEKs a saturation of the polymeric chain with thermo-resistant simple ether bonds is observed; on the other hand, the part of non steady complex ether bonds, which is brought into the structure by the remnants of dichloranhydrates of phtalic acids, falls sharply. Besides, the density of PEEK packing increases considerably with the rise of initial olygoketone length. These three facts in total may promote such a natural increase of heat-resistance of PEEK in this line.

The study of dielectric properties of PEEKs showed that the indices of dielectric penetration (table 1) in the lines fall a little (from 3,22 to 2,97 and from 3,43 to 3,06 correspondingly) with the increase of the initial OK length, and this may be explained by the PEEKs structure and the increase of crystallite degree for PEEKs with longer OK.

The synthesized PEEKs do not differ considerably in fire-resistance and the significance of oxygen index of these polyethers lie in the interval of 30,5-32,5%, which gives an opportunity to affirm that the given PEEKs will not burn in the atmosphere.

The obtained PEEKs are examined in their resistance to aggressive environment (table 2).

Table 2.The dependence of mass change of the models on time exposition

ПЭЭК на основе:	Exposure time s $\cdot 10^5$	Weight variations (%)				
		H_2SO_4 10%	H_2SO_4 30%	HCl Conc. acid	NaOH 10%	NaOH 50%
OK-1D	0,864	0,34	0,24	0,67	0,91	0,34
	1,728	0,98	0,90	1,68	2,64	-0,60
	3,456	1,61	1,76	2,71	2,69	-1,97
	13,824	1,69	1,34	2,88	-0,13	-10,13
OK-5D	0,864	0,29	0,27	0,70	0,80	0,30
	1,728	0,64	0,83	1,55	2,33	-0,87
	3,456	1,37	1,55	2,47	2,47	-2,16
	13,824	1,40	1,43	2,60	0,34	-9,91
OK-10D	0,864	0,30	0,17	0,58	0,81	0,24
	1,728	0,55	0,77	1,50	1,94	-0,79
	3,456	1,25	1,31	2,13	2,13	-1,81
	13,824	1,29	1,30	2,21	1,07	-7,64
OK-20D	0,864	0,11	0,15	0,33	0,62	0,12
	1,728	0,38	0,40	1,11	1,55	-0,54
	3,456	0,64	0,98	1,69	1,71	-0,97
	13,824	0,69	1,09	1.71	1,56	-6,33
OK-1F	0,864	0,72	0,87	0,74	0,44	0,57
	1,728	1,34	1,30	1,75	2,00	-0,66
	3,456	2,64	2,88	3,17	-0,09	-2,90
	13,824	2,69	2,60	3,20	-1,34	-17,63
OK-5F	0,864	0,60	0,80	0,91	0,66	0,70
	1,728	1,29	1,14	1,66	2,31	-0,13
	3,456	2,47	2,73	2,98	-0,31	-3,34
	13,824	2,54	2,60	2,95	-1,57	-15,49
OK-10F	0,864	0,43	0,61	0,80	0,50	0,72
	1,728	1,13	1,10	1,55	2,13	-0,17
	3,456	2,00	1,97	2,58	0,11	-0,20
	13,824	2,05	1,99	2,60	-0,94	-10,98
OK-20F	0,864	0,24	0,50	0,71	0,63	0,34
	1,728	0,95	1,01	1,65	1,79	-0,63
	3,456	1,53	1,46	2,00	1,37	-0,28
	13,824	1,58	1,51	2,01	-0,07	-8,88

As the results show, PEEKs on the basis of dian OK demonstrate a good resistance in diluted solutions of sulphuric acid and also in concentrated hydrochloric acid. They dissolve easily both in concentrated and diluted alkali which is connected with the presence of chemically unstable complex ether bonds in PEEKs chain. PEEKs on the basis of short OK and PEEKs, saturated with these bonds destruct faster than PEEKs on the basis of OK-20D. Besides, a low density of PEEK packing on the basis of short olygoketones promotes such speed of polymer destruction.

In 10%- sulphuric acid stabilisation of mass follows the swelling which witnesses of dissolvency absence and, what is more, the absence of destrucive processes. In 30%- sulphuric acid some loss of mass of the model on the basis of OK is observed after swelling. The examining of these models in viscidity showed that this characteristic remains without change which shows PEEK dissolvency but not the destructive process. A big swelling of the models is observed in a concentrated HCI, which is connected to the activity chlorous hudrogen steam.

The above-stated properties of polyetherketones on the basis of different olygoketones and dichloranhydrate of iso- and terephtal acids mixture show that these polymers, both pure and in different compositional materials on their basis may have a wide application as heat- and chemically resistant polymer materials for constructive purpose.

References

[1] Сайто, Тэруо. *Eng. Mater.* 1982, 30, №9. P 32-34.

[2] Иида, Хираси. Polyetherketones of Victrex. *Eng. Mater.* 1983, 31, №6. P 31-35.

[3] May, R. *Victrex aromatic polymers for high temperature application.* «Proc. 7th Annu Des. Ing. Conf., Birmingham, 25-27 Sept., 1984» Kempston, 1984. P. 313-318.

[4] Shinyama, K., Fujita, S. Dielectric and thermal properties of irradiated polyetheretherketone IEEE Trans. *Dielec. and Elec. Insul.*. 2001. 8, N 3, c. 538-542.

[5] Hamciuc, Corneliu, Bruma, Maria, Klapper, Marcus Sulfonated poly(ether-ketone)s containing hexafluoroisopropylidene groups *J. Macromol. Sci. A.* 2001. 38, N 7, P. 659-671.

[6] Reimer, W., *Weidig R. PEEK ersetzt Metall Kunststoffe.* 1996. 86, N 4, c. 528.

[7] High-temperature engineering polymers. *Pt II High-Tech. Mater.* Alert.1991. 8, N12, c. 5-6.

[8] *Patent №1736128* (USSR)

[9] *Patent №595889* (USSR)

[10] *Ramlow Gerhard Hochleistungskunststoffe Kunststoffe.* 2001. 91, N 9, P. 257-258.

[11] Shinyama, K., Fujita, S. Dielectric and thermal properties of irradiated polyetheretherketone IEEE Trans. *Dielec. and Elec. Insul.*. 2001. 8, N 3, c. 538-542.

[12] Sharapov, V.V., Shaposhnikova, V.V., Salaskin, S.N. Influence of conditions of polycondensation on synthesis polyarylenetherketones. *High-molecular connections.* 45. №1. P. 113-116.

In: Polymers, Polymer Blends, Polymer Composites... ISBN 1-60021-168-2
Eds: A.K. Mikitaev, M.K. Ligidov et al., pp.121-134

Chapter 15

POLYMERIC NANOCOMPOSITES, STABILIZED ORGANIC DERIVATIVES OF FIVE-VALENT PHOSPHORUS

A. Kh. Shaov**, *Kh. Kh. Gurdaliev and A. M. Kharaev
Kabardino-Balkarian State University, Nalchik, KBR, Russia

ABSTRACT

On the basis of aromatic polyesters and the block-copolyesters, and also polyethylene of high density compositions c were prepared the contents of organic compounds of five-valent phosphorus with cyclohexyl a radical at atom of phosphorus. The sizes of molecules phosphor organic compounds make 8-10 $Å^3$, that allows them to relate to so-called nanoparticles. It is established, that investigated phosphorus organic compounds appeared effective stabilizers thermal and mechanical properties of the investigated polymers, and also good antipyrens.

Key words: aromatic polyesters; polyolefin's; phosphor organic compounds; stabilization; nanoparticles.

The basic directions of researches in the field of increase thermo stability and fire resistance of polymeric materials are replacement of antioxidizers and antipyrens from compounds of bromine (and in general halogens) and threehidrate ammonium on more effective and less carcinogenic substances. Growing cost of compounds of antimony also has resulted in necessity of their replacement on phosphorus containing antioxidizers and antipyrens which in a combination to compounds of bromine, provide synergetic effect, that is obviously expressed in oxygenic polymers.

The general tendency in the given area of a science are questions of compatibility of additives with polymers, their influence on coloring of materials, shock durability and

*Kabardino-Balkarian State University, 360004, Nalchik, Chernyshevsky st. 173, KBR, Russia. Fax: (095) 337-99-55; E-mail: ah_shaov@mail.ru

adaptability to manufacture, and also development of target additives for concrete types of polymeric materials.

Nanocomposites a new class of the polymeric systems having extraordinary properties. The term "nanocomposites" means biphasic materials in which filled it is allocated in a polymeric matrix on nanolevel (10^{-9} m). Some more years back, before occurrence of the term "nanocomposites", the effects notice in polymeric compositions, additives of the most various origin containing small amounts, explained so-called «effect of small additives».

In the present work results of the researches received at studying of polymeric materials, containing some organic compounds are discussed on the basis of five-valent phosphorus in small concentration.

For an estimation of efficiency of stabilizers of polymeric materials use chemical, physical and chemical and physical-mechanical methods of research, but as most objective of them and important in the practical relation are considered physical-mechanical.

In this compounds, one axis stretching at speed of 40 mm/mines at 20^0 C us were investigated by a method mechanical properties of film samples of polymeric compositions. They are prepared by a method pour from a solution.

Polymeric compositions on a basis the block-copolyesters contain phosphor organic compounds (PhOC) in quantity of 0,5-30,0 % from weight of polymer, in case of a polycarbonate - in quantity of 0,1-30,0 %, and in PEHD - 0,1-0,5 %.

For an estimation of influence phosphor organic compounds on mechanical characteristics of polymers were used the following parameters: a breaking point (σ_b), the maximal relative lengthening (ε_b), the module of elasticity (E), and also impact strength (A_{st}) which are designed by known techniques. Influence phosphinic acids on mechanical characteristics the block-copolyesters can be looked after on change of values of explosive durability and the maximal relative lengthening depending on concentration phosphinic acids. It is established, that the greatest explosive durability among compositions from block-copolyester on a basis bisphenol A (BCP-7D) the structure containing 1,0 % phosphinic of an acid has. The further increase of concentration PhOC results in gradual decrease of breaking strength.

For block-copolyester on a basis phenolphthalein (BCP-7F) the maximal value of explosive durability falls to a composition with the contents of 10,0 % phosphinic to acids and is equal 139,0 MPa against 116,0 MPa for initial polymer. The greatest value of the maximal relative lengthening for compositions on basis BCP-7D also falls to structure with the contents of 1,0 % phosphinic to acids. For a composition on a basis phenolphthalein the block-copolyester this maximum is observed at the contents of 0,5 % of a researched acid.

The minimal value of the module of elasticity for compositions on the basis of both block-copolyesters falls to the contents phosphor organic to compounds in small quantities. It speaks that such structures are more elastic, than initial polymers.

With the purpose of finding - out of the mechanism of strengthening influence phosphinic acids at its use in rather small amounts on mechanical properties the block-copolyesters was carried out roentgen-structural the analysis of compositions on a basis diphenylolpropane polymer with the contents of various quantities phosphinic acids. Diffractogrames were removed on $Cu_{k\alpha}$ radiation (λ=1,5405 Å) on device DRON-3. It was established, that on diffractogrames both initial polymer, and the compositions containing various quantities phosphinic of an acid, it is not observed crystal areas.

Proceeding from experimental results, it is possible to assume, that small quantities phosphinic acids form the most thin monomolecular layers in interstructural areas macromolecules polymers which carry out a role so-called boundary greasing. In compositions there is "colloids" a system in which PhOC plays a role falling, i.e. phosphinic the acid represents itself as some softener of interstructural action. Besides since in a molecule phosphororganic compounds are available polar P=O and P-OH - groups it, carrying out a role falling, will participate in formation of intermolecular connections with polymeric groupings polymer macromolecules. The this interaction, the is stronger, in view of role PhOC as surface-activity compounds (SAC), probably, mobility over molecules is higher than structures meaning that orientation is facilitated at a stretching, that always promotes hardening of polymer. The further increase of concentration phosphinic acids, apparently, because of stereo the difficulties caused by presence in compositions of a plenty of rather voluminous molecules PhOC, results in easing intermolecular connections in polymers therefore the composite material collapses at smaller loadings. As to compositions on a basis phenolphthalein the block-copolyester and phosphinic acids when the maximal value of a breaking point falls to 10 % the contents phosphororganic to compounds in this case, apparently, for formation of a "continuous" most thin monomolecular layer, i.e. the boundary greasing facilitating mobility over molecules of structures, it is required a lot of surface-active substance (phosphinic acids) because of less dense packing polymer macromolecule on a basis phenolphthalein.

On curve dependences of a pressure on relative lengthening for compositions on a basis diphenylolpropane polymer also at the contents of small quantities phosphinic acids there is a displacement of curves in area of high values σ_b. At transition from phosphinic acids to her potassium salt character of influence phosphororganic compounds on mechanical properties the block-copolyester does not vary. Optimum concentration are small additives potassium phosphinate. But, as against the acid, at her use potassium salts of value of the module of elasticity in all a concentration interval are lower, and the maximal relative lengthening is higher, than in a case phosphinic acids. It is possible to explain greater polarity P-O-K - bond in a molecule phosphinate, in comparison with P-O-H - bond in a molecule of the acid.

With potassium salt in comparison with the structures containing phosphinic an acid it is possible to explain some decrease of durability of compositions larger in volume of an ion potassium, than atom of hydrogen in hydroxyl to group of an acid therefore arise additional stereo difficulties. It also it is possible to explain and some increase of values of the maximal relative lengthening for the compositions containing potassium phosphinate, i.e. because of increase of volume PhOC and, apparently, its polarity, force of intermolecular interaction between molecules potassium phosphinate as SAC and polar groupings macromolecules the block-copolyesters operate on larger distances in comparison with phosphinic an acid.

Use of nickel salt phosphinic acids as the additive to the block-copolyesters brings the contribution, though and not so distinguished from other compounds of five-valent phosphorus. It, first of all, is appreciable by consideration of influence nickel phosphinate on explosive durability of compositions. For diphenylol a block-copolymer of appreciable changes, as against the compositions containing phosphinic an acid or her potassium salt, it is not observed. For phenolphthalein the block-copolyester almost in all a concentration interval value of a breaking point is lower, than at initial polymer. It, most likely, is connected to the greater rigidity macromolecules itself phenolphthalein a block-copolymer in comparison with diphenylolpropane, and also stereo the difficulties brought by phosphinate molecules of nickel because of increase of number cycloalkyl and butoxyphenyl of groups as against phosphinic of an acid or potassium phosphinate.

As well as in a case potassium phosphinate, nickel salt phosphinic acids in small concentration raises value of the maximal relative lengthening, and the module of elasticity in all a concentration interval is lower, than size E of initial polymers.

Modifying influence nickel phosphinate on aromatic the block-copolyesters, as well as in a case with others PhOC, is more shown in relation to plastic deformation of polymers.

Phosphinate bivalent iron the block-copolyesters practically influences on mechanical properties the same as and nickel phosphinate, except for the module of elasticity - he starts to grow at the big concentration iron (II) phosphinate.

Such identity of character of influence nickel phosphinate and bivalent iron, apparently, is connected to identical assistants at cations both bivalent metals and approximately equal polarity of these molecules.

Character of influence phosphinate trivalent iron on mechanical properties the block-copolyesters practically reflects influence of other salts of bivalent metals: the tendency to reduction of explosive durability of compositions on a basis phenolphthalein a block-copolymer is evidently shown at increase of number phosphinate radicals in phosphor organically salts, that in turn depends on valency of metal. As to compositions on a basis bisphenol A the block-copolyester and iron (III) phosphinate here mechanical parameters remain on enough high level.

Research of influence phosphinic and its salts on mechanical properties of two poly-(arylatsulfonoxides) block-copolymers has shown acids, that for all phosphororganic compounds by the optimal concentration their small quantities, basically up to 3,0 % are.

In view of that processing polymers is improved, and at small quantities PhOC also are considerably improved mechanical characteristics of compositions on a basis the block-copolyesters and investigated derivative five-valent phosphorus, it is possible to assume, that offered compositions can find application as high-strength constructional polymeric materials.

After change of character of influence PhOC on mechanical properties the block-copolyesters was considered at transition from phosphinic acids to its salts, changing it the quantity phosphinate groups in a molecule of derivative five-valent phosphorus, was of interest research of influence on mechanical properties of polymers of replacement hydroxyl groups in a molecule phosphine acids on the second butoxyphenyl a grouping. Appeared, that at small concentration phosphinoxyde his action on mechanical properties the block-copolyesters comes nearer to those at use as modifiers phosphinic acids and her salts.

With increase of the contents phosphinoxyde in compositions on the basis of aromatic the block-copolyesters is shown the tendency to decrease of the investigated parameters. As well as others phosphororganic compounds, phosphinoxyde is more effective in relation to BCP-7D.

In a case phenolphthalein polymer, already since 1,0 % additives phosphinoxyde, plastic deformation starts to be reduced, but mechanical characteristics remain a little bit above, than with initial polymer. The further increase of concentration phosphinoxyde in compositions results in fragile destruction of samples.

In conclusion of consideration mechanical properties of compositions on the basis of aromatic block-copolyesters and phosphinic acids and her salts, and also phosphinoxyde, it is possible to notice the block-copolyesters, that all PhOC at their use in small concentration are the effective additives raising mechanical characteristics of investigated polymers. Such strengthening mechanical properties influence on polymers is shown more poorly in case of compositions on a basis phenolphthalein a block-copolymer. It first of all is connected to distinction in above molecules structures the block-copolyesters, caused by presence of the rests of molecules phenolphthalein in polymer on his basis.

In connection with that the stabilization considered as objects aromatic the block-copolyesters are only laboratory development was of interest to study, as the compounds of five-valent phosphorus received during researches will influence physical and chemical properties of industrial polyesters. As one of such polymers we choose a polycarbonate on a basis bisphenol A marks of the PC-3. It is established, that phosphinic the acid at her introduction in a polycarbonate does not render such appreciable strengthening influence on mechanical the characteristic of polymer as it took place in case of compositions on a basis the block-copolyesters.

On σ_b phosphinic the acid influences explosive durability of a polycarbonate practically a little though at her low concentration in compositions some increase of values of a breaking point is observed. Values of the maximal relative lengthening of compositions raise in comparison with initial polymer at use phosphinic acids up to 0,5 % a little. So, relative lengthening of a composition of the PC-3+0,5 % of phosphinic on 4,8 % is more than acid, than at an initial polycarbonate. The module of elasticity of compositions also is a little bit higher in comparison with the initial PC.

As the polycarbonate concerns to in part crystal polymers was of interest to find out influence PhOC on a degree crystallite a polycarbonate.

With this purpose in the same conditions, as for compositions on basis BCP-7D, was carried out X-ray structural the analysis polycarbonate compositions with the contents of various quantities phosphinic acids.

Compositions prepared as films a method pour from a solution in chloride methylene. It is established, that introduction phosphinic in small quantities results acids in some increase of a degree crystallite a polycarbonate. The increase of the contents phosphor organic up to 10,0 % appreciably reduces compounds crystallite polymer. The further increase of concentration practically does not render appreciable influence on a degree crystallite polymer. It is established, that on diffractogrames both a clean polycarbonate, and the compositions containing various quantities phosphinic of an acid, the crystal peak is on 17^0, that corresponds to the intermolecular distance equal 5,21 Å.

Proceeding from the received data it is possible to assume, that rather ineffective influence phosphinic acids on mechanical characteristics of a polycarbonate is connected to some infringement of orderliness of crystal packing macromolecules in a polycarbonate caused stereo by difficulties, voluminous molecules brought in compositions phosphororganic compounds.

It is known, that forces of an intermolecular attraction which bring the certain contribution in mechanical characteristics of polymers, operate on distance 3-4 Å. As in our case the intermolecular distance appeared little bit more than this parameter also the intermolecular attraction will be reduced a little, that, apparently, and results, along with stereo factors, to some downturn of strengthening influence phosphororganic compounds on a polycarbonate, in comparison with compositions on the basis of aromatic the block-copolyesters.

It is revealed, that at small concentration phosphinic acids plastic deformation and explosive durability of a polycarbonate remain on a high level, but the further increase of the contents phosphororganic results compounds in fragile destruction of samples.

Influence potassium phoshinate on mechanical properties of a polycarbonate, as well as in case of compositions on a basis the block-copolyesters, is more sharply expressed in comparison with influence of the acid.

The module of elasticity of compositions at the contents in them up to 3,0 % potassium phosphinate remains to higher in comparison with initial polymer, but further the module of elasticity of compositions is lower, than at a polycarbonate. Apparently, it is connected to increase of polarity phosphororganic compounds at transition from an acid to her potassium salts. Small additives potassium phosphinate raise plastic and mechanical characteristics of a polycarbonate a little.

Character of influence phosphinates bivalent metals (nickel and iron) on mechanical properties of a polycarbonate is practically identical: with increase of concentration of salts value of a breaking point tends to decrease, and the maximal relative lengthening raises at small quantities phosphinates a little; the module of elasticity of both systems gradually raises. Such identity of character of influence of these two salts on mechanical properties of a polycarbonate, probably, is connected to an identical structure of molecules phosphinates.

Phosphinate trivalent iron monotonously reduces explosive durability of a polycarbonate, and value of relative lengthening at break for the contents of salt in quantity of 5,0 % practically does not vary, but further is reduced. The module of elasticity of salt at small concentration varies also a little, and at increase of quantity phosphinate iron (III) this parameter grows.

Introduction phosphinoxyde in a polycarbonate reduces mechanical characteristics of polymer a little. It is possible to explain it to that given phosphororganic compounds less, apparently, as least polar of all investigated compounds of phosphorus, forms (in comparison with phosphinic an acid and her salts) intermolecular bonds with macromolecules a polycarbonate, but at the same time molecules phosphinoxyde have rather big volume and they can represent itself as loosening macromolecules some polymer of the agent therefore the intermolecular distance is a little increased, as results in some decrease of mechanical properties of compositions in comparison with initial polymer.

In connection with that in the industry the stabilized variant of a polycarbonate is issued, and as the stabilizer against destruction of thermo oxidation apply derivative trivalent phosphorus - three(p-nonylphenyl)phosphit (Polygard), was of interest to study, as influences Polygard on mechanical properties of a polycarbonate and to compare him to character of influence investigated by us phosphor organic compounds. With this purpose, and also for revealing an opportunity of display synergetic effect at simultaneous presence at compositions Polygard and synthesized phosphororganic compounds, we investigated mechanical properties of the industrial stabilized polycarbonate with the contents of 1,0 % Polygard, and the compositions containing derivative trivalent phosphorus and investigated compounds of five-valent phosphorus in quantity of 0,1-30,0 %. It is established, that the stabilized variant of a polycarbonate containing Polygard, has the worse parameters mechanical characteristics in comparison with not stabilized and modified offered PhOC (Table. 1).

Table 1.Mechanical properties of compositions on the basis of a polycarbonate and phosphor organic compounds

No	Composition	σ_b, MPa	ε_b, %	E, GPa
1	Polycarbonate (PC)	88	17,6	1,6
2	PC + 1,0 % Polygard	82	14,7	2,3
3	PC + 0,5 % phosphinic acids	108	34,4	2,0
4	PC + 1,0 % potassium phosphinate	101	62,1	1,5
5	PC + 3,0 % nickel phosphinate	104	56,4	1,5
6	PC + 10,0 % phosphinate Fe (II)	103	51,6	1,6
7	PC + 0,1 % phosphinate Fe (III)	155	91,4	1,8
8	PC + 1,0 % phosphinoxyde	94	14,7	2,1

Besides apparently from tab.1, small concentration phosphororganic compounds render appreciable improving effect at their sharing with Polygard. The highest parameter of mechanical properties from all investigated compositions with phosphororganic compounds has trivalent iron phosphinate at his use in quantity of 0,1 %.

Thus, comparison of compositions on the basis of the industrial polycarbonate, containing simultaneously offered by us phosphororganic compounds on the basis of five-valent phosphorus and industrial stabilizer Polygard, allows to approve, that composite materials with improved mechanical properties are received.

In connection with that the stabilization used as objects aromatic the block-copolyesters have high impact strength (in particular for BSP-7D it not less than 117-118 kJ/m^2), were of interest to determine character of influence received phosphororganic compounds on impact strength the block-copolyesters. With this purpose compositions on a basis bisphenol A a block-copolymer and phosphororganic compounds in quantity of 1,0 % from weight of polymer were prepared. The choice of such concentration phosphororganic compounds is caused by the optimal contents of these compound by consideration quasi static mechanical properties of compositions. It is established, that investigated phosphororganic acids reduce impact strength of polymer. The highest values of impact strength were received for compositions with the contents) of salts of iron and phosphinic acids. It is interesting to note, that among salts the most effective appeared derivative cyclohexyl phosphonic acids. So, the composition with the contents nickel hydrocyclohexylphosphonate has shown value A_{st}=107 kJ/m^2, with the contents bivalent iron hydrocyclohexylphosphonate - 115 kJ/m^2, and compositions with trivalent iron cyclohexylphosphonate had A_{st}=136 kJ/m^2.

To one of the major factors determining resistance of polymers to shock loadings, plasticity of polymeric materials and improvement of this characteristic of polymers concerns will promote increase of shock durability of samples. Told practically proves to be true the data received as a result of research of influence PhOC on impact strength the block-copolyester on a basis diphenylolpropane, and also industrial polyarilate DV- salts are more effective modifiers by way of improvement of shock durability and as it was shown above, plasticity of polymers.

The difference in the mechanism of influence phosphororganic compounds on mechanical behavior of polymers, most likely, is connected from them above molecule by structure. Block-copolyester BSP-7F it is worse "diphenylol" than a block-copolymer on mechanical to characteristics for this reason.

It is necessary to allocate the following factors influencing properties of a composition. Except for above molecule structures (AMS) of initial polymer, it is chemical structure PhOC, its volume, a degree of polarity and a way of distribution of the additives- modifiers, changes of defectiveness influencing a degree AMS of polymer. Besides it is necessary to take into account that circumstance, that the mentioned above factors can render various influence on mechanical behavior of a composition at different stages of deformation.

In case BSP-7D (to a lesser degree BSP-7F, meaning) introduction PhOC in small quantities results initial parameters to hardening a block-copolymer, is increased σ_b and ε_b. The module of elasticity simultaneously decreases. At an explanation of such character of change of mechanical properties it is necessary to mean, that these characteristics of polymers are determined by the total contribution as inside, and inter chain interaction. Their parity depends and is determined as actually by deformation (size ε), and a temperature-time mode mechanical loading. In conditions quasi static the tests used in the present work, speed of deformation is sufficient smaller ($\sim 10^{-2}$-10^{-4} c^{-1}) to consider, as inside, and inter chain interaction "have completely time" to realize the influence on mechanical behavior of polymer. The above efficiency of intermolecular interaction, the more its contribution to size of the module of elasticity and rigidity as a whole a composition at all stages of deformation. We shall specify, that the module of elasticity is determined on an initial site of the diagram "σ-ε" at very small values of lengthening. For this reason the assumption of role PhOC as SAC and «boundary greasing» is quite allowable. The second mechanism - strengthening of intermolecular interaction as each of considered phosphororganic compounds to some extent is active in this sense in parallel operates.

In view of small contents PhOC and, all the same, reduction of size E, it is logical to assume, that in this concentration interval and in the field of deformations where the module of elasticity is determined, the contribution to its formation of the first mechanism prevails. In process of development of deformation «boundary greasing» facilitates reorganization AMS of a block-copolymer resulting in its steadier to destruction to structure. At the deformations close to ε_b, on durability all mentioned above factors have to the full an effect. Influence of polarity PhOC in this case is evidently shown: the maximal values of a breaking point decrease in process of its growth. The same dependence σ_b and from volume of molecules phosphor organic compounds. These experimental facts confirm the assumption that at small contents PhOC the mechanism of interstructural plasticization the block-copolyesters prevails. Comparing BSP-7D and BSP-7F, influence and mechanisms of influence on mechanical behavior BSP-7F of small quantities PhOC more ambiguous is possible to come to a conclusion, that.

Decrease with increase of contents PhOC mechanical parameters the block-copolyesters speaks redistribution of the relative contribution of plasticization and strengthening of intermolecular interaction (IMI) both in durability, and in the module of elasticity. Thus it is important to note, that phosphinic an acid and phosphinoxyde at high contents render more effective complex influence on mechanical behavior of compositions: rather high mechanical parameters and growth of the module of elasticity. General, though and in a different degree expressed, result - reduction ε_b, and influence on parameters of plasticity of polymer.

Loosen structures of blocks-copolymers can occur at increase of contents PhOC on various mechanisms. However it is obvious, that efficiency of the additive now in the greater degree is determined, except for structural features, volumetric characteristics (steric the factor), activity phosphor organic compounds in relation to IMI. In this respect higher efficiency of some PhOC can be caused by an optimum parity of parameters of the additive and distribution of free volume in polymer. It is possible to assume, that the sizes appropriate of ~150-250 sm^3/mol – Van der Waals to volume (V_w) phosphinic acids and phosphinoxyde are such. Similar assumptions can be considered and concerning polarity phosphor organic compounds. That strengthening of efficiency IMI ambiguously influences on mechanical properties of compositions that is visible from comparison of some compositions on basis BSP-7D and BSP-7F proves to be true also.

In a polycarbonate the free volume is less, than in the block-copolyesters. On the other hand is in part crystal polymer with rather crystal high degree. In this case again it is necessary to find an optimum parity of properties of additive PhOC (for example, its volume) with distribution of free volume in polymer. Besides the polycarbonate is: more loosen packing and denser packing areas of an amorphous part, crystal and inter crystal areas. Therefore is unequivocal to determine the mechanism of influence PhOC on properties of a polycarbonate more difficulty. For example, the size dense packing areas of the PC is estimated by different authors, different experimental methods in limits 30÷100 Å besides, that its volume fraction of ~50 %. On the other hand molecules phosphor organic additives ($V_w \approx$ 8-10 $Å^3$) borrow, most likely, free volume in more loosen packing parts of amorphous area and, with smaller probability, inter crystal areas.

In view of told above, it was possible to expect less effective strengthening influence of all phosphor organic compounds on a polycarbonate, and laws at his studying can be others, is especial regarding influence of salts of metals phosphinic acids. Here the combination of small concentration and properties PhOC (for example,) has given some salt of trivalent iron the best result.

It is necessary to note, that shock tests differ that in conditions of high-speed deformation the relative contribution to durability and plasticity of effective intermolecular interaction sharply grows. Here it is necessary to take into account that impact strength A_{st} - the integrated power characteristic of durability of polymer or a composition, showing power expenses for destruction, σ_b and E - power.

Results of shock tests (A_{st}) in very big degree depend on plasticity of samples. On the other hand, in plasticity brings the contribution not all free volume, but only his "effective" part. Efficiency of influence of additives PhOC on shock durability depends, thus, on a parity of the sizes of a molecule of the additive with "effective" free volume of polymer and, thus, from as far as concentration PhOC is picked optimum up. In a case block-copolyester BSP-7D in conditions of shock tests are salts of iron and phosphinic acids; for the PC-trivalent iron phosphinate.

One of the basic operational characteristics of polymeric materials is shock durability in this connection structures on the basis of polyethylene of high density (PEHD) are investigated by a technique of shock tests on Sharpy.

Experimental installations with which help polymeric systems are investigated represents pendulum koper UT-1/4, supplied by the piezoelectric gauge of loading, the signal with which moved directly on remembering oscillograph of model from C 8-12. Character of influence PhOC on physical-mechanical properties PEHD in conditions shock loading estimated under such characteristics, as A_{st}, E, ε_b, σ_b, σ_{ce} and ε_{ce} which values are calculated with use of settlement formulas: $A_{st} = W_{st}/B(D\text{-}a)$; $E = (P\times L^3)/4\delta B\,(D\text{-}a)^3$; $\varepsilon = [6\delta(D\text{-}a)]/L2$; $\sigma_b = 3P_{st}L/2B(D\text{-}a)^2$; $\sigma_{ce} = 3PbL/2B(D\text{-}a)^2$, where A_{st}- impact strength; W_{st}- energy of destruction of a sample; B- width of a sample; D- thickness of a sample; and - size of a cut; P - loading on an initial linear site of the diagram; δ- a deflection of a sample in its middle; L- distance between support of a pendulum; ε_b- deformation of destruction; σ_{ce}- a limit of the compelled elasticity; ε_{ce}- deformation of the compelled elasticity; σ_b- a breaking point; P_{ce}- loading on diagram P-t (time) appropriate to macroscopically compelled elasticity; P_{st}- loading of destruction.

The received results testify that phosphinic an acid and her salts render plastically influence on polyethylene of high density. To draw such conclusion some increase of impact strength allows at simultaneous decrease of the module of elasticity and increase of relative deformation at destruction of samples.

All phosphor organic compounds appeared practically more effective additives to polyethylene in comparison with the industrial stabilizer polyolefin's Irganox-1010.

At the same time it is necessary to note, that phosphinic acid salts reduce a breaking point whereas the acid practically does not render essential influence on the given parameter a little.

Character of change A_{st} at introduction in polymer PhOC correlates with change of values of a pressure of the compelled elasticity. The given fact can speak about some increase of intermolecular interaction in polymer, that, apparently, is connected by that phosphor organic compounds, occupying "free" volumes in macro chain, the polar groupings strengthen intermolecular interaction a little.

Such behavior of modifiers should result in some improvement of orderliness in macro chain, that practically and occurs, if it to judge on increase of a degree of crystallinity (α_m) (Tables 2, 5). Thus the degree of crystallinity raises with transition from an acid to its salts, i.e. with increase of polarity phosphor organic the modifier.

By consideration of character of influence cyclohexyl phosphonic acids, her potassium salts, and also phosphine oxide on physical-mechanical properties PEHD in conditions shock loading it is noticed (Tables 3, 4), that the behavior of data PhOC practically does not differ from those phosphinic acids and its salts.

The results received at research of character of influence phosphor organic of compounds on physical-mechanical characteristics PEHD in conditions shock loading, allow to assert with the big share of reliability, that irrespective of molecular weight PhOC show plastic property, i.e. the mechanism of influence of modifiers, probably, does not vary. It is necessary to note, that salts phosphinic acids reduce value σ_b polyethylene whereas the acid practically does not render essential influence on the given parameter a little. Character of change of impact strength PEHD at introduction in him PhOC correlates with change of values of a pressure of the compelled elasticity that can speak about some increase of intermolecular interaction in polymer. It, apparently, is connected by that phosphor organic compounds occupying free volumes in macro chain, the polar groups strengthen intermolecular interaction a little. As confirmation of such assumption that circumstance can serve, that at Van der Waals volume (V_W) polyethylene in 20,6 sm^3/mol, found on a known technique the share of free volume (V_E) makes 7,6 sm^3/mol and shares V_W phosphor organic compounds, their falling investigated dosages (0,05-0,5 %), make only 0,005-0,088 sm^3/100 г polymer. The effect of small additives, probably, is connected by that at such dosages PhOC they in the optimum image "find room" in free volume of polymer, though own Van der Waalsoves volumes phosphor organic compounds are greater (58,1-423,0 sm^3/mol), than at polyethylene. For the industrial stabilizer polyolefin's Irganox-1010 value V_W even above also makes 711,1 sm^3/100 г polyethylene.

Table 2. Influence phosphor organic modifiers on density and degree of crystallinity PEHD

No	Composition	d, g/sm^3	α_m
1	PEHD	0,951	0,701
2	PEHD+0,1 % Irganox-1010	0,958	0,746
3	PEHD+0,05 % phosphine acids	0,952	0,707
4	- « - + 0,1 % - «-	0,957	0,739
5	- « - + 0,3 % - «-	0,951	0,701
6	- « - + 0,5 % - «-	0,957	0,739
7	PEHD+0,1 % - « - + 0,1 % Irganox-1010	0,957	0,739
8	PEHD + 0,05 % nickel phosphinate	0,952	0,707
9	- « - + 0,1 % - «-	0,959	0,752
10	- « - + 0,3 % - «-	0,955	0,727
11	- « - + 0,5 % - «-	0,957	0,739
12	PEHD+0,1 %-«-+ 0,1 % Irganox-1010	0,955	0,727
13	PEHD + 0,05 % iron (II) phosphinate	0,958	0,746
14	- « - + 0,1 % - «-	0,954	0,720
15	- « - + 0,3 % - «-	0,956	0,733
16	PEHD + 0,05 % iron (III) phosphinate	0,957	0,739
17	- « - + 0,1 % - «-	0,960	0,758
18	- « - + 0,3 % - «-	0,952	0,707
19	- « - + 0,5 % - «-	0,958	0,746

Note: the degree of crystallinity α_m is designed proceeding from density

Table 3. Physical-mechanical properties of compositions on a basis PEHD and PhOC in conditions of shock test

No	Composition	A_{st}, kJ/m^2	E, GPa	σ_b, MPa	ε_b, %	σ_{ce}, MPa	ε_{ce}, %
1	PEHD	11,0	1,06	21,1	5,7	27,9	4,8
2	PEHD+0,1 % Irganox-1010	12,8	1,55	18,6	6,4	30,0	3,6
3	PEHD+0,05 % phosphinic acids	22,6	0,81	19,9	7,9	29,1	4,7
4	- « - + 0,1 % - «-	18,5	0,79	17,8	8,3	29,3	4,6
5	- « - + 0,3 % - «-	18,8	0,79	18,1	9,1	29,7	5,1
6	- « - + 0,5 % - «-	11,5	0,94	20,2	6,4	28,9	4,4
7	PEHD+0,05 % nickel phosphinate	14,6	0,85	15,9	9,0	25,5	4,2
8	- « - + 0,1 % - «-	17,9	0,89	14,1	10,6	26,9	4,6
9	- « - + 0,3 % - «-	22,1	0,85	12,8	8,3	26,9	4,7
10	- « - + 0,5 % - «-	17,4	0,72	17,5	7,9	27,1	5,3
11	PEHD+0,05 % iron (II) phosphinate	12,8	0,85	16,0	7,9	28,7	4,3
12	- « - + 0,1 % - «-	11,6	0,85	16,3	7,7	27,1	4,7
13	- « - + 0,3 % - «-	9,8	0,77	12,0	7,5	23,9	4,5
14	PEHD+0,05 % iron (III) phosphinate	13,3	0,77	11,2	8,4	23,9	4,7
15	- « - + 0,1 % - «-	17,6	0,82	15,9	8,1	24,7	4,5
16	- « - + 0,3 % - «-	14,4	0,78	12,5	9,4	21,0	4,7
17	- « - + 0,5 % - «-	6,4	0,62	13,9	8,3	23,0	4,9

It is necessary to note, that polyethylene - more "simple" structure, than the block-copolyesters and a polycarbonate, with the big factor of packing, besides in part - crystal. His studying is complicated that amorphous his phase at room temperatures no glassing, and at shock tests is necessary to bring artificial defect - a cut to have an opportunity supervision of the full diagram of destruction in coordinates "force-deformation".

Table 4. Physical-mechanical properties of compositions on a basis PEHD and PhOC in conditions of shock test

No	Composition	A_{st},KJ/m^2	E, GPa	σ_b, MPa	ε_b, %	σ_{ce}, MPa	ε_{ce}, %
1	PEHD	11,0	1,06	21,1	5,7	27,9	4,8
2	PEHD + 0,1 % Irganox-1010	12,8	1,55	18,6	6,4	30,0	3,6
3	- « - +0,05 % phosphinic acid	19,5	0,82	14,1	9,5	28,1	5,0
4	- « - + 0,1 % - «-	16,3	0,90	18,6	7,6	28,5	4,6
5	- « - + 0,3 % - «-	18,8	0,81	17,1	7,6	28,5	4,7
6	- « - + 0,5 % - «-	20,0	0,87	16,5	8,7	28,5	4,8
7	- « - +0,1 %-«-+ 0,1% Irganox	22,6	0,85	16,2	8,3	28,7	4,8
8	PEHD+0,05 % potassium monophosphonate	20,0	1,05	29,9	5,1	30,6	3,9
9	- « - + 0,1 % - «-	19,3	0,85	16,0	8,1	27,9	4,9
10	- « - + 0,3 % - «-	15,0	0,82	16,0	7,5	27,1	4,8
11	- « - + 0,5 % - «-	12,1	0,57	18,7	6,9	27,3	4,7
12	- « - +0,1 %-«-+ 0,1 % Irganox	23,1	1,03	15,2	9,8	30,4	5,3
13	PEHD+0,05 % potassium phosphonate	20,5	0,92	16,0	8,6	25,4	4,1
14	- « - + 0,1 % - «-	19,0	0,85	18,0	7,6	26,7	4,6
15	- « + 0,3 % - «-	15,5	0,91	17,3	8,4	26,7	4,8
16	- « + 0,5 % - «-	15,4	0,82	18,3	8,5	27,9	4,6
17	- « + 0,1 %-«-+ 0,1 % Irganox	19,5	0,84	15,7	7,9	26,3	4,6
18	PEHD+0,05 % phosphinoxyde	21,8	0,90	12,8	8,8	27,1	4,6
19	- « - + 0,1 % - «-	20,7	0,90	13,3	9,5	28,5	4,7
20	- « + 0,3 % - «-	18,2	0,80	16,0	7,9	26,9	5,2
21	- « + 0,5 % - «-	19,7	0,85	18,3	7,9	27,1	4,4
22	- « + 0,1 %-«-+ 0,1 % Irganox	17,4	0,94	16,5	7,7	27,7	4,4

Therefore it is necessary to take into account not only structure, volume, properties of additives PhOC, but also as far as the put cut (~5 mm) is close or far from structural defect PE. By influence on σ_b, the critical border between PhOC in sense of their influence on defectiveness of a matrix passes between phosphinic an acid and her salts. The integrated estimation of influence PhOC on mechanical properties of polyethylene at impact results in the same conclusions, as earlier, in case of consideration the block-copolyesters and a polycarbonate.

Table 5. Influence cyclohexyl phosphonic acids and her derivativeson density and degree of crystallinity PEHD

No	Composition	d, g/sm^3	α_m
1	PEHD	0,951	0,701
2	PEHD + 0,1 % Irganox-1010	0,958	0,746
3	- « - + 0,05 % phosphonic acid	0,954	0,720
4	- « - + 0,1 % - «-	0,946	0,669
5	- « - + 0,3 % - «-	0,956	0,733
6	- « - + 0,5 % - «-	0,951	0,701
7	- « - + 0,1 % - « - + 0,1 % Irganox-1010	0,954	0,720
8	PEHD + 0,05 % potassium monophosphonate	0,957	0,739
9	- « - + 0,1 % - «-	0,958	0,746
10	- « - + 0,3 % - «-	0,951	0,701
11	- « - + 0,5 % - «-	0,957	0,739
12	- « - + 0,1 % + 0,1 % Irganox-1010	0,954	0,720
13	ПЭВП + 0,05 % potassium phosphonate	0,954	0,720
14	- « - + 0,1 % - «-	0,949	0,688
15	- « - + 0,3 % - «-	0,951	0,701
16	- « - + 0,5 % - «-	0,954	0,720
17	- « - + 0,1 % - « - + 0,1 % Irganox-1010	0,955	0,727
18	PEHD + 0,05 % phosphinoxyde	0,962	0,771
19	- « - + 0,1 % - «-	0,950	0,695
20	- « - + 0,3 % - «-	0,951	0,701
21	- « - + 0,5 % - «-	0,954	0,720
22	- « - + 0,1 % - « - + 0,1 % Irganox-1010	0,946	0,669

In the conclusion it is possible to ascertain, that high efficiency synthesized phosphor organic compounds in quality inhibitors thermo oxidation destruction aromatic the block-copolyesters is established - the appreciable effect is observed at the contents of these compounds in compositions in quantity up to 1,0 %.

On increase of stabilizing effect phosphor organic compounds form a line: p-butoxyphenyl cyclohexyl phosphinic an acid, her potassium salt, di-(p-butoxyphenyl) cyclohexylphosphinoxyde; higher efficiency of the synthesized new compounds of five-valent phosphorus - di-(p-butoxyphenyl) cyclohexylphosphinoxyde - is found out as the stabilizer for a polycarbonate in comparison with the traditional stabilizer for this polymer - Polygard, being derivative of trivalent phosphorus - [tris-(p-nonylphenyl) phosphit], that is shown, in particular, in the best ability to preservation of durability and elasticity after long tests of polymer at 423 K; It is established, that synthesized phosphor organic the compounds containing in molecules atoms of five-valent phosphorus, influence on polyethylene of the low pressure, similar plastifying to effect that is shown in increase of impact strength and relative lengthening at simultaneous decrease of the module of elasticity in conditions of shock tests.

REFERENCE

[1] Wilku, Ch. A. Polymer degradation and stabilization.-*Polym. News*. - 1995.-20.-No 10. - P.316-317.

[2] Schetlick, K., Habicher, W.D. Organophosphores antioxidants action mechanismes and new trends: Pap. 16 th Int. Conf. Adv. and Degrad. Polym. Luzern.-June 14-17.-1994.- *Angew. makromol. Chem*. - 1995. - 232. - P. 239-246.

[3] Foigt, I. Stabilization of synthetic polymers against action of light and heat.-L.: *Chemistry*. - 1972.-544 p.

[4] Shrainer, R., Fyuzon, R., Kertin, D., Moryll T. *Identification of organic compounds*. - *M*. - 1983. - 704 p.

[5] Ozden, S., Shaov, A.H., Charayev, A.M., Bidanikov A.Y. Compositions Based in Aromatic Block Copolyester and p-Bytoxyphenyl Cyclohexyl Phosphinic Acid.-*Polym. and polym. Comp*.-1998.-Vol. 6.-No 2.-P.103-107.

[6] Ozden, S., Shaov, A.H., Charaev, A.M., Gurdaliyev X.X. The effects of p-butoxyphenyl-cyclohexylphosphinic acid on the properties of PC based on bisphenol A. - *J. Appl. Polym. Sci*.-2001.-Vol. 80.-P. 2113 - 2119.

[7] Ozden, S., Shaov, A.H., Charayev, A.M., Mikitaev A.K., Bedanokov A.Y. Aromatic block copolyesters stabilised with metallic salts of phosphinic acid.-*Polymers and Polymer Composites*.-2001.-Vol. 9.-No 3.-P. 213-219.

[8] A.Kh. Shaov, N.N. Amerkhanova, A.M. Kharaev. Phosphorous Organic Compounds as the stabilizers of the Aromatic Blockpolyethers. - *Aging of polymers, polymer blends and polymer composites*.-Nova Sci. Publ.-New York.-2002.-Vol. 2.- P. 161-166.

[9] Shaov, A.Kh., Sarieva, Z.I., Gurdaliev, Kh. Kh. The Research of the Character of the Influence of Phosphinates and Phosphinoxyde on Mechanical Properties of Polyethylene High Density. - Nova Sci. Publ. - *Aging of polymers, polymer blends and polymer composites*. - New York. - 2002.-Vol. 2.-P. 167-170.

[10] Shaov, A.Kh., Dubayeva, E.Kh., Mikitaev, A. K. Stabilization of Polycarbonate with some Organic Derivatives of the Five-Valent Phosphorus. - *Aging of polymers, polymer blends and polymer composites*. - Nova Sci. Publ. - New York. - 2002.-Vol. 2.-P. 171-174.

[11] Shaov, A.Kh., Kodzokova, E.Kh. Organic derivative five-valent phosphorus as stabilizers and modifiers of polymeric materials (Review).-P.1 – *Plasticheskie Massy*. - 2004. - No 12. – P. 21-34.

[12] Shaov, A.Kh., Kodzokova, E.Kh. Organic derivative five-valent phosphorus as stabilizers and modifiers of polymeric materials (Review).-P.2. - *Plasticheskie Massy*. - 2005. - No 3. – P. 33-41.

[13] Shaov, A.Kh., Borukaev, T.A., Begretov, M.M. Last achievements in the field of creation of fire-resistant polymeric materials with use phosphor organics compounds (Review.) - *Chemical physics of pyrolysis, combustion and oxidation*.-Nova Sci. Publ., New York.-2005.-P.19-31.

In: Polymers, Polymer Blends, Polymer Composites… ISBN 1-60021-168-2
Eds: A.K. Mikitaev, M.K. Ligidov et al., pp.135-142

Chapter 16

POLYURETHANEISOCYANURATE POLYMERIC MATERIALS

L. V. Luchkina*, A. A. Askadskii, K. A. Bychko and V. V. Kazantseva
A.N. Nesmeyanov Institute of Organoelement Compounds,
Russian Academy of Sciences, Moscow, Russia

ABSTRACT

Polymeric materials based on Poly(oxypropyleneglycol), 2,4-toluenediisocyanate, and di-(3-chloro-4-aminophenyl)methane have been prepared. Materials mentioned above possess elastic behavior and its modulus of elasticity vary from 3.6 to 1250 MPa. It is evident that the polyurethaneisocyanurates prepared by us possess good physico-mechanical and physico-chemical.

Keywords: polyurethane isocyanurates, diisocyanate, isocyanurate ring, elasticity modulus, specific impact strength

Elastic polymeric materials derived from polyoxypropylene glycol, aromatic diisocyanates (MDI, TDI), and aromatic diamine possessing elasticity modulus in the range of 3.6 – 1,250 MPa were synthesized and studied [1 – 3]. The basis for the process of these networks synthesis is polycyclotrimerization of bifunctional monomer (diisocyanate) and an oligomer with end isocyanate groups. This reaction allows for formation of a network with trifunctional isocyanurate rings, produced in the interaction of three isocyanate groups:

* A.N. Nesmeyanov Institute of Organoelement Compounds, Russian Academy of Sciences, 119991, 28 Vavilov St., Moscow, Russia.lara.luch@mail.ru

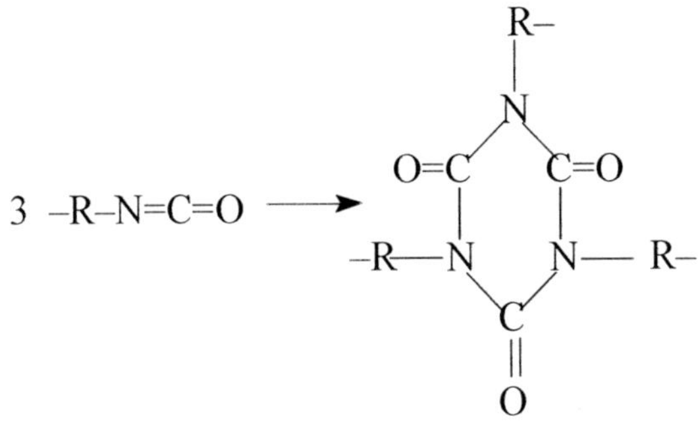

The reaction between macrodiisocyanate and aromatic diamine (di(3-chloro-4-aminophenyl)methane) was also used. Chemical structure of this network depends on the ratio between rigid bulky triisocyanurate crosslinked points and flexible linear fragments linking them. The products have different properties with respect to the ratio between crosslinked points and flexible chains.

Oligomeric diisocyanate (macrodiisocyanate) OEC was synthesized according to technique described in [2]. The progress of the reaction was controlled using IR-spectroscopy method by the band intensity at 2,280 cm^{-1} and the bands of hydroxyl groups at 3,500 – 3,600 cm^{-1} [4]. The quantity of isocyanurate groups (%) was estimated by their reaction with dibutylamine, which excess was titrated by an acid [5]. Resulting the reaction, formation of urethane groups absorbing in the range of 3,300 – 3,400 cm^{-1} was determined from IR-spectrum [6]. OEC was synthesized as a viscous colorless liquid (at room temperature); $n_D^{30} = 1.4711$, $\rho = 1.057$ $g{\cdot}cm^{-3}$.

For synthesizing polyurethane-isocyanurate networks possessing a wide controllable range of properties the methods of polycyclotrimerization and migration polymerization were applied. At the second stage of the process, OEC produced was mixed with variable TDI quantity, added with DMBA and ED-22 based amino-epoxide catalyst and variable quantity of diamine. As a result of catalytic interaction of isocyanate groups which enter into the composition of OEC and TDI, a 3D polyisocyanurate network structure representing a transparent gel, insoluble in acetone, was formed [2].

In its turn, MDC interaction with diamine first produced linear polyurethane urea, which was confirmed by spectroscopic data on the band intensity at 1,550 cm^{-1} [6, 7]:

$$\mathrm{OCN \sim\sim NCO + H_2N{-}R{-}NH_2 + OCN \sim\sim NCO + H_2N{-}R{-}NH_2}$$
$$\downarrow$$
$$\mathrm{{-}OC{-}NH \sim\sim NH{-}CO{-}NH{-}R{-}NH{-}CO{-}NH \sim\sim NH{-}CO{-}NH{-}R{-}NH{-}} \qquad (1)$$

where ~~~ is macrodiisocyanate molecule.

Excessive MDC not entering into the reaction induced the chain crosslinking. At such crosslinking and temperature increase above 100°C MDC was added by preliminarily formed uric bonds, which are reactive enough in relation to diisocyanate groups. As a result, one more bond type – biuretic, occurred in crosslinked rubbers:

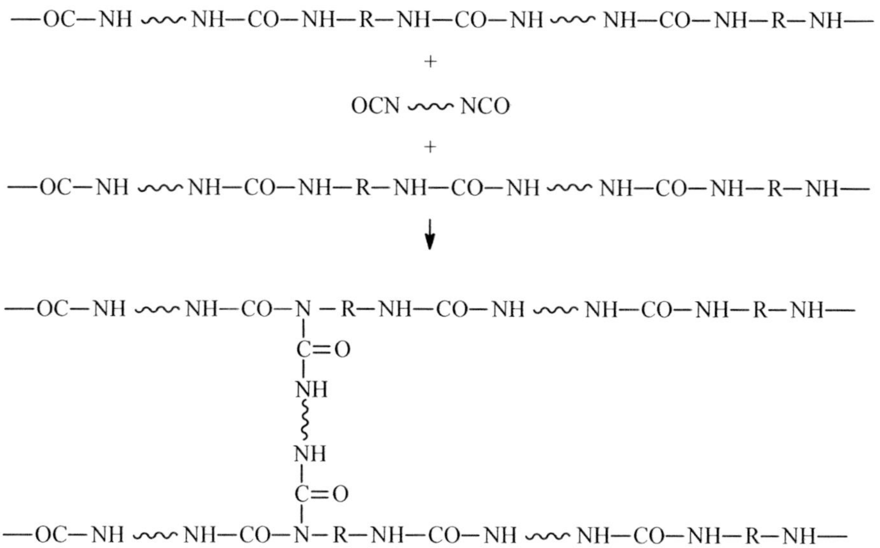

(2)

It is found that in spectra of polymers synthesized the bonds are present, typical of allophanate structures (1,310 cm^{-1}) [6]. This testifies about high probability of the reaction between urethane group and isocyanate at temperature increase to 120 – 140°C during synthesis of polyurethane isocyanurate materials under selected conditions.

The reactions of urea interaction with isocyanate and urethane group with isocyanate are very important for synthesis of polyurethane isocyanurate materials, because they induce formation of branched and crosslinked structures in polyurethanes. As is known, these structures are hard-wearing. These reactions are slowly proceeding (during 5 – 6 h), but are suitable because of required temperature level, equal to polyisocyanurate synthesis (120 – 140°C), also usually applied in the industry.

The behavior of samples derived from polyurethane isocyanurate networks under conditions of thermomechanical and TGA analyses were considered. Figure *1a* shows thermomechanical curves of seven samples containing different quantity of polyurethane (PU) component and TDI, also differing by the elasticity modulus values.

As found in previous observations [8], polyisocyanurate networks display two transitions with respect to their composition: the first is low-temperature transition associated with devitrification of the rubbery phase, and high-temperature related to devitrification of the glassy phase. Hence, both transitions are temperature–shifted towards one another, and the progress of such shift depends on the microphase composition. Therefore, Figure *1b* presents separately thermomechanical curves of polyurethane isocyanurate samples related to the low-temperature area.

For a network polymer composed only by PU or containing the greater quantity of it (80 wt.%) (Figure *1b*, curves *1 and 2*), an increase of low-temperature transition temperature is insignificant, because the rubbery phase contains a small quantity of TDI residues. Vice versa, in a plastic sample with lower PU content or its full absence and consequently higher

TDI content, such shift is rather intensive. As a result, the low-temperature transition for the system containing no TDI is displayed at ~ –50°C, whereas for the sample with 60 wt.% TDI (Figure *1b*, curve *7*) is occurs at +50°C. Concerning the high-temperature transition, for the low-modular sample (Figure *1*, curves *1, 2*) it is observed at 150 – 190°C. Meanwhile, as TDI content in the sample increases (the content of polyisocyanurate – PIC – component increases), the transition happens at higher temperatures of about 320 – 330°C.

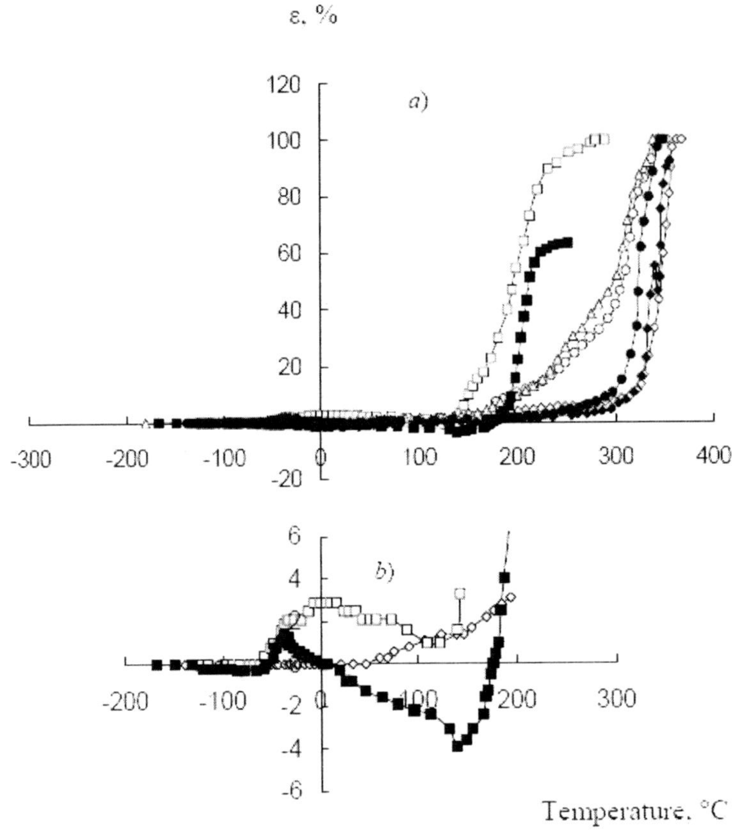

Figure 1. Temperature dependence of deformation of polyurethane isocyanurate polymeric materials. The material contains PU composites in amount of 100(1), 80 (2), 60 (3), 50 (4), 40 (5), 20 (6), and 0 wt.% (7)

The comparative analysis of thermodynamic (Figure *1*) and TGA curves (Figure *2*) shows that for all samples, the initiation temperature of intensive thermal degradation falls in the range of 320 – 330°C. This means that for low-modular samples, high-temperature transition observed at 180 – 190°C is not related to thermal degradation of the polymer. On the contrary, for high-modular samples, this transition is observed at the temperature, at which intensive thermal degradation proceeds. Therefore, it may be suggested that deformability of synthesized high-modular network systems is associated with degradation processes, proceeding in these systems at high temperatures. Therefore, in the systems such as high-modular polyurethane isocyanurates, heat resistance is limited by their thermal stability.

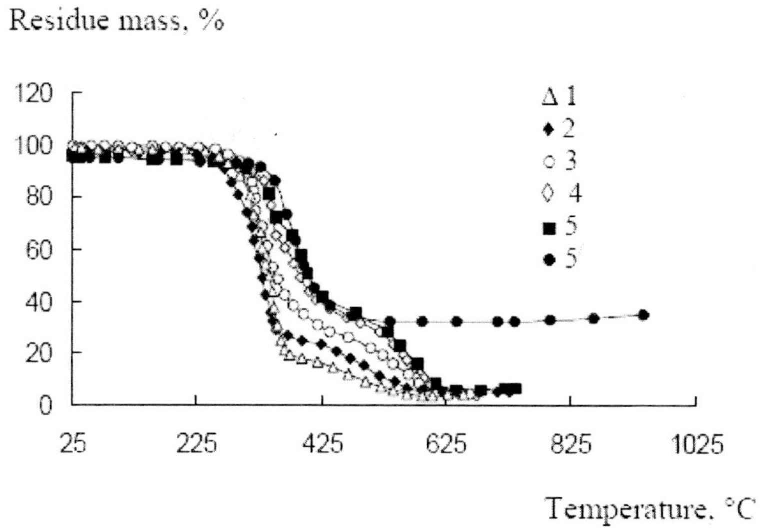

Figure 2. TGA curves for polyurethane isocyanurate polymeric materials in air (1 – 5) and in argon (5′). The material contains the following quantities of PU composites: 100 (1), 80 (2), 50 (3), 20 (4), and 0 wt.% (5, 5′)

The increase of PIC concentration in the composite raises thermal stability of the polymer so higher, the higher TDI concentration is (Figure *2*, curves *1, 5*). Comparing with PU containing no TDI, the PIC composite with 60% TDI possesses temperature on intensive thermal degradation by 40°C higher. In the inert medium (argon), temperatures initiating weight decrease and intensive PIC degradation are not increased compared with degradation in the air (Figure *2*, curves *5, 5′*). However, degradation in argon gives much higher coke residue.

Thus, according to TGA and TMA data., it is possible to estimate the temperature interval, in which polyurethane isocyanurate polymers may operate without significant softening and degradation. For low-modulus polyurethane-isocyanurates, this range runs from –50 to 190°C, and for high-modulus composites it may reach 330°C.

We have also studied the influence of the PU/PIC ratio on physicomechanical properties of synthesized polyurethane isocyanurates. As would be expected (Table *1*, n. *6*), the highest specific impact viscosity (up to 12 kg×cm/cm^2) is typical of a polymer, in which isocyanurate crosslink points of the network are linked by flexible polyurethane networks containing propylene oxide groups. Therefore, the higher PU concentration and, correspondingly, the lower TDI concentration is in relation to OEC in the initial composite, the higher specific impact viscosity is, or the material is not degraded at all, but bends during tests on "Dynstat" device. However, such indices as bending strength and elasticity modulus reach their minima. In its turn, the elasticity modulus reaches its maximum in the polymer with the maximal PIC concentration, which contains the maximal quantity of aromatic diisocyanate (Table *1*, n. *1–3*). Injection of different fillers to initial composites for the purpose of synthesizing polyurethane isocyanurate polymeric materials leads to changes in values of elasticity modulus, specific impact viscosity, and bending strength. For example, injection of 5 wt.% TiO_2 at 50 wt.% TDI concentration if the initial reaction mixture (Table *1*, n. *2*), specific impact viscosity is 5-fold increased. However, in the absence of TiO_2 and at 60 wt.% TDI concentration specific impact viscosity is much lower (Table *1*, n. *1*). The injection of technical carbon in amount of 2 wt.% also 2-fold increases specific impact viscosity of the samples. The injection of the above-mentioned additives also increases elasticity modulus of the samples.

Table 1. Physicomechanical properties of polyurethane isocyanurate polymeric materials

No	PIC:PU, wt.%	Elasticity modulus (*E*), MPa	TDI content in the mixture with OEC in PIC composite, wt.%	Specific impact viscosity, kg·cm/cm^2	Bending strength, kg/cm^2
1.	PIC (without filler)	1185	60.0	1.04	771
2.	PIC (with TiO_2)	1250	50.0	5.6	718.3
3.	PIC (with technical carbon)	1314	60.0	2.0	240.0
4.	80:20	517	48.0	3.9	232.5
5.	60:40	293	36.0	8.9	183.8
6.	50:50	127	30,0	12.0	30.0
7.	40:60	21.7	24.0	Does not degrade	27.8
8.	20:80	14.4	12.0	Does not degrade	12.8
9.	PU	3.9	0	Does not degrade	10.8

Note that polyurethane isocyanurate samples containing a great PU quantity and having elasticity modulus value falling in the range between polymeric glass and rubber, display elastic rather than viscoelastic behavior, which is typical of all known polymers present in the transition zone from the glassy to the rubbery state. In this connection, to analyze the type of mechanical behavior of synthesized polyurethane isocyanurate materials, relaxation curves for different values of elasticity modulus were calculated. The stress relaxation curves were composed in relative stress – time coordinates. The relative stress was calculated as σ_i/σ_0, where σ_i is the current relaxing stress; σ_0 is the initial stress developed at the end moment of deformation setting. As observed from the Figure, currently synthesized polymeric materials show an abrupt stress reduction at the initial stage of relaxation with further transition to extremely low stress relaxation. It is also important that the composition of current materials is significant for the progress of relaxation curves of relative stresses, σ_i/σ_0.

The estimation of water effect on synthesized polymeric materials (Table *2*) indicated insignificantly higher water absorption and water absorption rate of low-modular polyurethane isocyanurate polymeric materials compared with high-modular ones. The oil resistance of high-modular samples reaches 0.15%. This value is so higher the higher PU concentration in the initial composite for the synthesis of polyurethane isocyanurates is.

Table 2. Physicochemical properties of polyurethane isocyanurate polymeric materials

No	PIC:PU wt.%	TDI content in the mixture with OEC in PIC composite, wt.%	Water absorption capacity ($B\cdot10^4$), g/dm^2	Water absorption rate ($W\cdot10^4$), g/dm^2	Oil resistance during 24 h, %	Fire resistance
1.	100	50	2.2	2.09	0.15	*
2.	80:20	40.0	2.35	2.12	0.34	The same
3.	60:40	30.0	2.40	2.27	0.65	The same
4.	50:50	25.0	2.67	2.45	1.6	The same
5.	40:60	20.0	2.73	2.55	2.85	**
6.	20:80	10.0	2.89	2.83	40	The same
7.	100	0	3.03	2.92	80	The same

* The sample weakly combusts longer than ¼ minute still.

** The sample highly combusts longer than ¼ minute still.

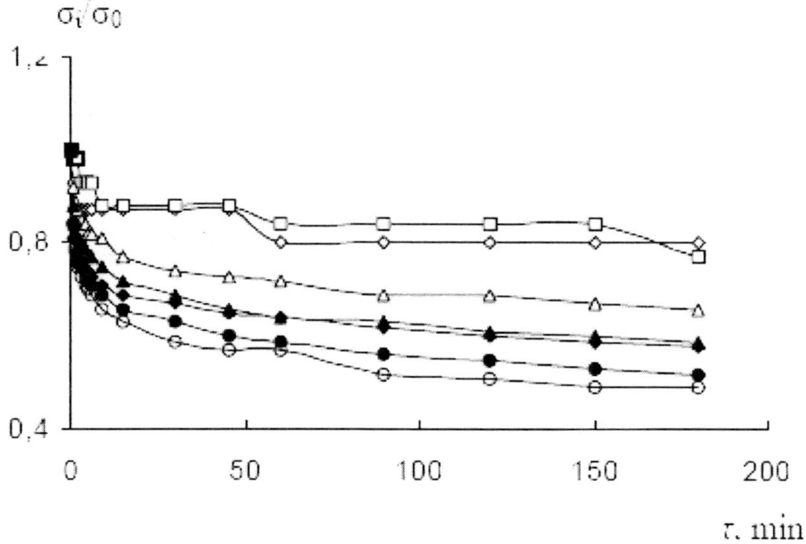

Figure 3. Relative stress relaxation curves for polyurethane isocyanurate polymeric materials. The materials contains PU composite in amounts: 100 (1), 80 (2), 60 (3), 50 (4), 40 (5), 20 (6), and 0 wt.% (7)

Finally, oil resistance of the elastic part reached 80%. After 24 h exposure to an oil product, low-modulus samples swell and crack. Fire resistance tests carried out by the technique described in [9, 10] showed their combustion time over 1/4 min.

Thus, thermal and heat resistance, as well as physicomechanical and physicochemical properties of high-modulus and low-modulus polyurethane isocyanurates, synthesized by OEC and TDI polycyclotrimerization and migration MDC polymerization with diamine were studied. The influence of the ratio of these components on the mentioned properties is also studied. It is shown that polyurethane isocyanurates synthesized in the selective reactions of polycyclotrimerization and polymerization possess quite high physicomechanical and physicochemical properties.

REFERENCES

[1] Askadski, A.A., Luchkina, L.V., Bychko, K.A., Goleneva, L.M., and Konstantinov, K.V., *Vysokomol. Soed.*, 2004, vol. 46A(4), p. 569. (Rus)

[2] Askadski, A.A., Luchkina, L.V., Bychko, K.A., Goleneva, L.M., and Konstantinov, K.V., *Vysokomol. Soed.*, 2005, vol. 47A(5), p. 763. (Rus)

[3] Askadski, A.A., Luchkina, L.V., Goleneva, L.M., Kiseleva, T.I., and Bychko, K.A., *Electronic Journal "Studied in Russia"*, 2004, vol. 69, pp. 751 – 769. (Rus) http: // zhurnal. ape. relarn. ru/ articles/ 2004 / 069. pdf.

[4] Bellamy, L., *Infrared Spectra Of Complex Molecules*, 1963, Moscow, Inostr. Lit. (Rus)

[5] *Polymer Encyclopedia*, 1972, vol. 1, Moscow, Sovetskaya Entsiklopedia, p. 1224. (Rus)

[6] Right, P. and Cummings, A., *Polyurethane Covers*, 1973, Leningrad, Khimia. (Rus)

[7] Lipatov, Yu.S., Kercha, Yu.Yu., and Sergeeva, L.M., *The Structure And Properties Of Polyurethanes*, 1970, Kiev, Naukova Dumka. (Rus)

[8] Askadski, A.A. and Kondrashchenko, V.I., *Computerized Material Science Of Polymers*, 1999, Moscow, Nauchnyi Mir. (Rus)

[9] Grigoriev, A.P. and Fedotov, O.Ya., *The Technology Of Plastics*, 1977, Moscow, Vysshaa Shkola. (Rus)

[10] Arkhangelski, B.A., *Plastics*, 1961, Leningrad, Gos. Soyuz. Izd. Sudostr. Prom. (Rus)

In: Polymers, Polymer Blends, Polymer Composites... ISBN 1-60021-168-2
Eds: A.K. Mikitaev, M.K. Ligidov et al., pp.143-149

Chapter 17

THE ESTIMATION OF OPPORTUNITIES OF LOW-TEMPERATURE DESTRUCTIONS OF SYNTHETIC RUBBERS IN SOLUTIONS IN RECEPTION OF HALF-FINISHED PRODUCT FOR FINISHING COMPOSITIONS

L. L. Kovalevskaja** and A. M. Ivanov
Kursk State Technical University

ABSTRACT

The opportunity and expediency of use of catalytic and oxidizing destruction of synthetic rubbers in solutions reactionary compounds in quality of half-finished product are confirmed at reception with getting several finishing compositions. Many factors of influence on oxidative-destructive transformations of the various marks of rubbers, favorable for realizations of the specified opportunity are identified and characterized.

Keywords: Rubbers, Solutions, Reactionary Compounds,Catalytic Destruction, Oxidizing Destruction in Solutions, Finishing Compositions, Factors of Influence, Reception, Management of Process.

The oxidative-destructive transformations of synthetic rubbers (SR) in solutions allow to receive reactionary compounds with various qualitative and quantitative functional structures of products (Table. 1).

It has been found [1,2], that the structure of a final reactionary compound is a direct function of many conditions of carrying out oxidative -destructive transformations. In particular:

* Kursk State Technical University, fhht@pochta.ru

- the nature of rubber (natural, butadiene rubber, styrene-butadiene rubber (SBR), methylstyrene-butadiene rubber (MSBR), isoprene rubber (IR), butyl rubber (BR), etc.);
- the nature of the solvent used (white-spirit, solvent, xylene, spirits, a mix of spirits and ketons);
- the conditions of preliminary reception of a solution and the contents of rubber in it (in a range of 5-15 % mass);
- the carrying out a full or partial dissolution SR directly in a reactor for oxidative-destructive transformations;
- the degrees of overlapping of dissolution SR and its oxidative-destructive transformations, and also duration of such a period;
- the nature, quantity and a phase condition of the catalyst or specially picked up catalytic systems;
- the type of the reactor used, the hydrodynamic conditions in it and the dynamics of change of the last in the course of the process;
- a temperature mode of a process carrying out (in most cases not isothermal);
- the organizations and quantitative characteristics of air bubbling or other kinds of contact of a reactionary compound with it;
- a choice of the moment of the oxidative-destructive process discontinuance and carrying out stabilization of the reactionary compound obtained;
- of some other individual factors in particular cases.

Table 1. The ranges of contents of the nonlimiting and oxygen-containing products in the final reactionary compounds, received at destructive oxidation SRS-30 in white-spirit at temperature 90 ÷ 105 °C

Connections	A range of content in mole/kg of a initial firm rubber	A range of concentration in mole/kg of a reactionary compound
Peroxide	0,20÷0,50	0,01÷0,08
Nonlimiting	3,00÷23,00	0,25÷2,76
Carbonyl	0,50÷4,30	0,05÷0,45
Spirits	0,50÷9,00	0,05÷1,40
Acids	0,20÷0,68	0,01÷0,08
Epoxy	0,14÷0,23	0,02÷0,03
Hydrolyze	0,60÷8,80	0,03÷1,25

The examples of kinetic curves of change of a liquid phase viscosity and accumulation of the separate products of transformation are shown in Fig. 1 and 2. It is well seen, that the process mostly begins with some, sometimes a small induction period, removable by the period of the autoaccelerated development, achievement of the maximal speed and the subsequent fading. Thus as a rule there is a maximum on the kinetic curves of accumulation of functional groups, which time characteristics for separate functional groups are different. It allows to use time of the discontinuance of oxidative -destructive process as an important factor of management of the received products structure.

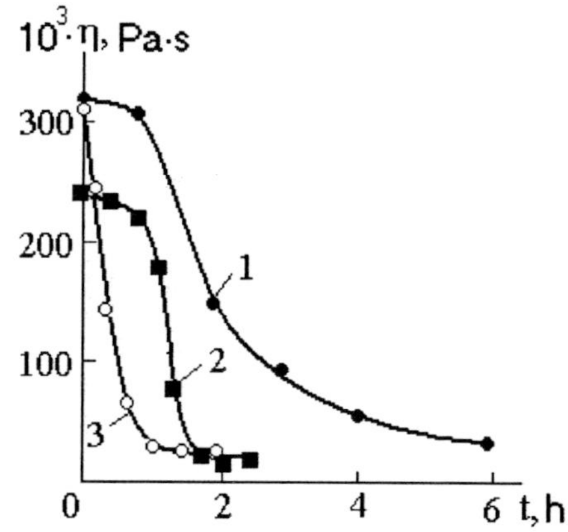

Fig. 1. Kinetic curve of changes of a reaction-ary compound dynamic viscosity at oxidative-destructive transformations (1,3) and cataly-tic destruction (2) of 13,7 % and 13 % solu-tions of SBR-30 in white-spirit at 100 °C ina reactor with pneumatic (1) and mechanical (2,3) hashing (1440 rev/min)

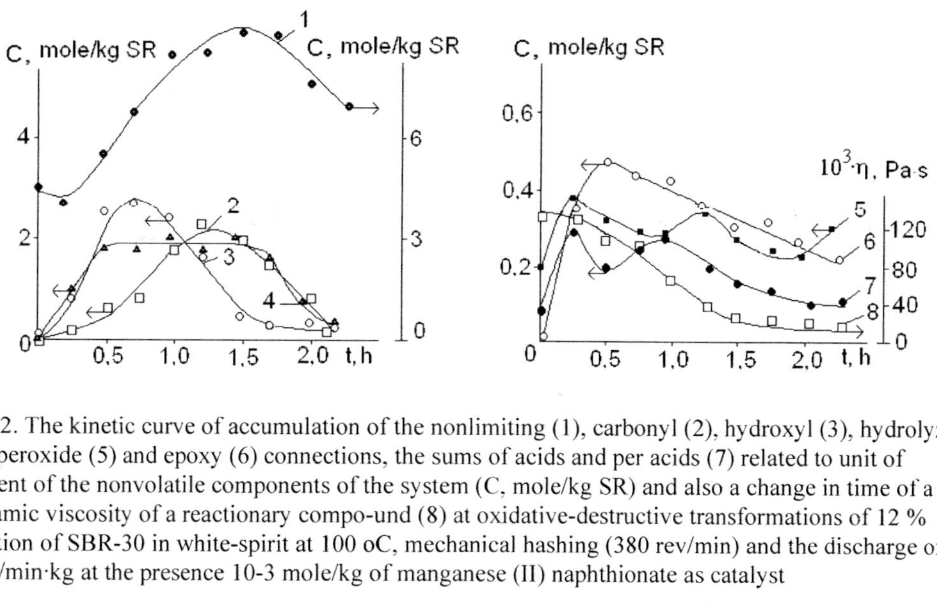

Fig. 2. The kinetic curve of accumulation of the nonlimiting (1), carbonyl (2), hydroxyl (3), hydrolyze (4), peroxide (5) and epoxy (6) connections, the sums of acids and per acids (7) related to unit of content of the nonvolatile components of the system (C, mole/kg SR) and also a change in time of a dynamic viscosity of a reactionary compo-und (8) at oxidative-destructive transformations of 12 % solution of SBR-30 in white-spirit at 100 oC, mechanical hashing (380 rev/min) and the discharge of air 0,2 l/min·kg at the presence 10-3 mole/kg of manganese (II) naphthionate as catalyst

The oxidative -destructive transformations, the liquid-phase oxidation of olefins, the oxidizing and catalytic destructions proceed in accordance with chain radical mechanism with origin of chains with the participation of the catalyst. One of variants of the latter is the reaction between the manganese-containing heterogeneous catalyst and olefins, having trailer double bond [1]:

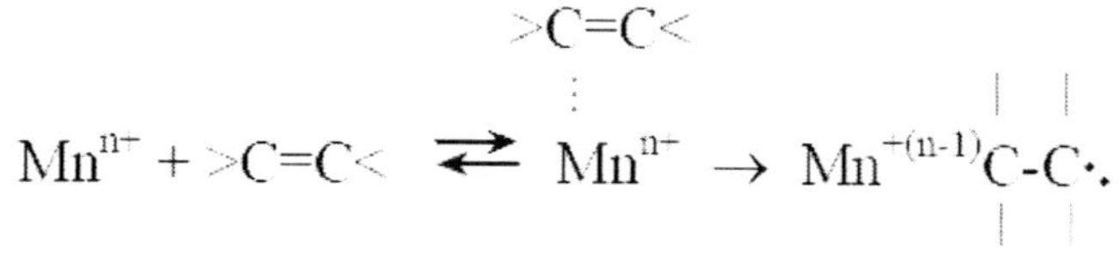

The following values of speeds of chains origin have been received by a method of inhibitors for catalytic destruction of 7 % solution of SBR-30 in white-spirit at 95 °C at the presence of 10^{-3} mole/kg of manganese (II) salts in a metal reactor at mechanical hashing (1440 rev/min) [3]: $(0{,}6\div4{,}0)\cdot10^{-6}$ mole/l.s (for resinate at use quinone and hydroquinone); $(0{,}8\div5{,}0)\cdot10^{-6}$ mole/l·s (for naphthionate at use quinone and hydroquinone); $(0{,}4\div5{,}0)\cdot10^{-6}$ mole/l·s (for benzoate at use quinone and hydroquinone); and for catalytic destruction of 10 % a solution of butyl rubber (BR) in white-spirit in the same conditions: $4{,}4\cdot10^{-6}$ mole/l·s (for stearate manganese at use quinone). The following values of speeds of chains origin have been received for oxidative-destructive transformations of 10 % solution of SBR-30 in white-spirit at 90 °C at the presence of 10^{-3} mole/kg of manganese (II) salts in a metal reactor with rabble (380 rev/min) and the charge of air 0,25 l/min·kg: $(1{,}4\div2{,}6)\cdot10^{-6}$ mole/l·s (for resinate at use quinone); $(2{,}6\div4{,}2)\cdot10^{-6}$ mole/l·s (for resinate at use hydroquinone); $(1{,}4\div2{,}0)\cdot10^{-6}$ mole/l·s (for resinate at use of iodine); $(2{,}0\div3{,}4)\cdot10^{-6}$ mole/l·s (for benzoate at use quinone); $(2{,}8\div4{,}0)\cdot10^{-6}$ mole/l·s (for benzoate at use hydroquinone); $(2{,}0\div2{,}5)\cdot10^{-6}$ mole/l·s (for benzoate at use of iodine). They are in good conformity with the data, available in the literature for modeling reactions, considered type [4-6].

The optimum interval of temperatures for the considered oxidative-destructive transformations is 90÷105 °C. The growth of viscosity of a solution affects at lower temperatures, that results in different reduction of process speed. At higher temperatures (120÷130 °C) the catalyst loses activity as a result the intensive system dehydration with partial loss by salts of their crystallized waters, and also as a result the transition of the basic salts in average, in particular as a consequence of oxidation of system reducers by manganese (III) and (IV) salts. The waterless manganese (II) salt dissolved better in tens times. In a result more often more active in considered systems the heterogeneous catalyst of destruction disappears and concentration of the homogeneous catalyst more effective already for polymerization processes grows.

Qualitative and quantitative comparison of the kinetic characteristics of the processes of catalytic destruction butyl rubber (BR), SBR-30, butadiene rubber and isoprene rubber (IR) in solutions of white-spirit at the presence of manganese (II) salts has been carried out [3]. Many analogies take place in the qualitative plan, but quantitative characteristics are predetermined by nature SR in the greater degree. SBR-30, butadiene rubber, IR destroy easiest. Destruction BR, which is characterized by the least compared by the contents in a polymeric molecule of isolated multiple bonds, proceeds with the least speed.

In the course of process of catalytic destruction SR in solutions there are strong changes of molecular-mass distributions of destruction products, that through respective alterations of structure of a reactionary compound affects its physical characteristics. In most cases kinetics of viscosity changes has been submitted monotonously decreasing curve (fig. 1), though there are variants linear (SBR at rather low quantities of the catalyst), step (BR) and even extreme (with a minimum) character. It is a consequence of that in considered system processes of condensation develop alongside with destruction, and competing ability of these chemical transformations in the course of process can vary in very wide limits. In case of catalytic oxidative-destructive transformations there are even more essential changes in functional structure of reactionary compounds as a result of the parallel with destruction courses of liquid-phase oxidation of destruction products.

Influence of the nature and quantity of the catalyst, and also concentration SR in an initial solution on duration of process and structure of products in reactionary compound has been appreciated. Examples in this plan are submitted on Fig. 3 and in Table. 2 [1].

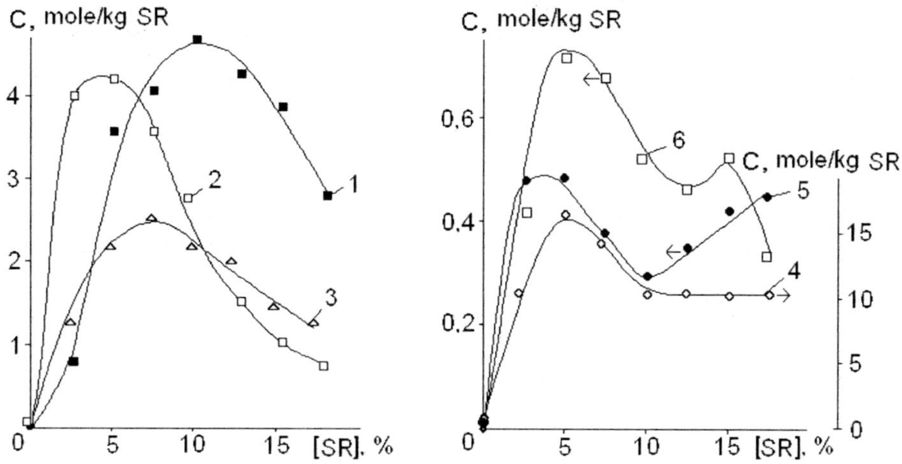

Fig. 3. The maximal yield of the hydroxyl (1), carbonyl (2), hydrolyze (3), non-limiting (4) and peroxide (5) connections, the sums of acids and per acids (6) related to unit of content of the nonvolatile components of the system (C, mole/kg SR) depen-ding on mass content of SBR-30 in initial loading at its oxidative-destructive transfor-mations into a solution of white-spirit at 100 °C, mechanical hashing (380 rev/min) and the discharge of air 0,33 l/min·kg at the presence 10^{-3} mole/kg of manganese (II) stearate as catalyst

Table 2. As many as possible achieved yields of the basic products of oxidative-destructive process in dependence upon nature used for creation catalytic systems of transitive metal salts

A salt of a transitive metal besieged from the water environment	A maximal yield of the products, mole/kg SR			
	Nonlimiting connections	Spirits	Carbonyl connections	Hydrolyze connections
manganese stearate	22	9,0	1,4	7,0
manganese resinate	18	5,1	1,1	7,3
manganese benzoate	20	2,8	4,0	7,6
manganese naphthionate	10,0	3,4	3,0	5,0
copper stearate	10,5	5,0	1,4	8,8
cobalt resinate	22	9,0	1,2	8,0
cobalt naphthionate	21,5	6,0	4,0	6,0
lead resinate	23	3,4	3,3	2,2

On the basis of results of preliminary executed research, some final reactionary compounds, representing solutions, a mix of the limited soluble in each other solutions or rather proof emulsions of products of destructive and oxidative-destructive transformations SR into hydrocarbonic solvents has been offered to use as target finishing compositions and half-finished product for them without additional operations concentration, divisions. Thus, in particular, technologies of reception have been developed:

- butyl-rubber drying oils and paints through a stage of destructive oxidation butyl rubber in a solution of white-spirit [7];
- oil-butyl-rubber drying drying oils and paints through a stage of joint oxidation at 95÷105 °C oxidized of butyl rubber, sunflower oil and terpene oils in the ratio on dry substance 1:0,5:1,5 [8];
- oil-rubber siccative-free drying oils, paints and enamels by one-phasic joint oxidation SR such as SBR, MSBR, butadiene rubber, IR in solvent with terpene oil and sunflower oil in the ratio on dry substance 1:1,5:0÷0,5 in an interval of temperatures 90÷105 °C [9];
- rubber varnishes for rubber footwear through a stage of catalytic destruction of SBR-30 solution in white-spirit in an interval of temperatures 90÷105 °C [10].

The operational experience gained has allowed to formulate recommendations on the storage of the obtained film-forming bases. The degree of destruction is one of the major factors, determining their safety. The reactionary compound representing a solution or a mix of solutions of products of transformation in hydrocarbonic solvents, in which are metal-containing the connections, capable to sedimentation at upholding, as suspension, are most stable. And the quantity gone into a solution metal-containing connections should be as small as possible, as further they favour development and course of processes of structurization. At big degrees of oxidative-destructive transformations the reactionary compound from a solution passes in emulsion, i.e. becomes a less stable system. Hence, it is possible to count a degree of destruction limited, predetermining transition of system from solution to emulsion. It means, that the duration of a process is the way of management not only functional structure and consumer properties, but also stability of obtained final reactionary compounds.

Strong deviations from an optimum mode of course of reception of film-forming bases processes result in reception of unstable compositions, while being stored inclined to spontaneous polymerization with capture of all solvent of system and rubbery compound formation.

On the whole the determined ways of management of destructive and oxidative-destructive processes have considerably facilitated the rational decision of a problem of processing of waste products of manufacture of mass purpose SR by reduction of their molecular mass and change of functional structure with the purpose of optimization of properties of received film-forming and others half-finished product for finishing compositions of a different special-purpose designation.

References

[1] Kovalevskaja, L.L. Kinetic regularity of destructive-oxidative transformations of styrene-butadiene rubbers in solutions in conditions of intensive mechanical hashing: *Author's abstract dissertation cand.chem.sci.:* 02.00.04. Kursk, 1996.-19 p.

[2] Burih, G.V. Research of kinetics and catalysis of low-temperature destructive-oxidative transformations of styrene-butadiene rubbers in conditions of pneumatic hashing of a liquid phase: *Author's abstract dissertation cand. Chem.Sci.:* 02.00.04. Kursk, 1997.-18 p.

[3] Kovalevskaja, L.L., Ivanov, A.M. Low-temperature destructions of sub-standard synthetic rubbers in solutions in reception of film-forming bases for finishing compositions // *New polymeric composite materials: materials II of the* All-Russia scientific-practical conference.-Nalchik: Kab.-Balk. Univ., 2005.-P. 262-266.

[4] Kozak, S.I., Nikipanchuk, M.V., Chernjak, B.I. Kinetic regularity of octene-1 liquid-phase oxidation on heterogeneous catalysts // *Petrochemistry*, 1976.-№ 5.-P. 740-743.

[5] Kinetic regularity of octene-1 liquid-phase oxidation at presence MnO / Kotur, M.G., Kozak, S.I., Nikipanchuk, M.V., Chernjak, B.I. // *Kinetics and catalysis*.-1988.-V. 29.-№ 5.-P. 1258-1261.

[6] Initial stages of octene-1 liquid-phase oxidation at presence MnO_2 / Nikipanchuk, M.V., Kotur, M.G., Kozak, S.I., Chernjak, B.I. // *Kinetics and catalysis*.-1986.-V. 27.-№ 5.-P. 1110-1114.

[7] Inventor's certificate. USSR №1728274 under the application №4685096/05 cl. C09D 123/26, C08C 19/04. *A way of reception of a basis for drying oil* /Ivanov, A.M., Rozanova, E.N., Ryzhkov, J.G., etc. BI, 1992, №15.-7 p.

[8] The patent of the Russian Federation №2026328; under the application №5007618 (05) cl. C09D 115/00, C09D 109/00, 193/00. *A way of reception of film-forming* / Ivanov, A.M., Kudryavtseva, T.N., Ivanova, L.A., etc. BI, 1995, №1.-8 p.

[9] The patent of the Russian Federation №2129580 under the application №96102447/04 cl. C09D 109/00 C08C 19/04. *A way of reception of film-forming* / Ivanov, A.M., Kovalevskaja, L.L., Ivanov, И.A. BI, 1999, №12.-10 p.

[10] The patent of the Russian Federation №2109752 under the application №95116561/04 cl. C08C 19/08; C09D 115/00. *A way of reception of a film-forming of black rubber varnish* / Ivanov, A.M., Kovalevskaja, L.L., Ivanov, И.A. BI, 1998, №12.-5 p.

In: Polymers, Polymer Blends, Polymer Composites… ISBN 1-60021-168-2
Eds: A.K. Mikitaev, M.K. Ligidov et al., pp.151-154

Chapter 18

TEMPERATURE TRANSITIONS IN POLYCARBONATE –POLYTETRAMETHYLENOXIDE BLOCK COPOLYMER RESINS

R.C. Bazheva, A.M. Kharayev, A.K. Mikitayev, G.B. Shustov and Z.L. Beslaneeva
Kabardino-Balkarian State University,
360004, Nalchik, Chernishevskaya St. 173, KBR, Russia

ABSTRACT

The main temperature transitions in polycarbonates and polycarbonate – polytetramethylenoxide block copolymer resins, obtained by acceptor-catalytic polycondensation in solution, were studied by the method of differential scanning calorimetry.

Key words: polycarbonate, polycarbonate –polytetramethylenoxide block copolymer resins, acceptor-catalytic polycondensation, vitrification temperature, melting temperature, crystallity.

Physical-chemical properties of polymers, and particularly block copolymer resins (BCR) are determined by their phase state and phase morphology to a significant degree [1]. That is why these characteristics of polymers are very important for understanding a number of processes, taking place in their treatment and use.

It is known that macromolecules of polycarbonate (PC) are characterized by a great rigidity, limited by the rotation of aromatic nuclear and due to this reason they are weakly tend to crystallization. Industrial PC is a vitreous polymer which has both short and long range ordering areas. The degree of crystallity does not usually overcome 10-15 %, and it can reach 30-40% only after special treatment of specimens [2].

Evaluation of physical structure of BCR is of a great interest. It is known from special literary resources that the character of phase separation and phase morphology in BCR depends on the chemical structure of blocks, molecular weight and the sequence of blocks in macromolecules, the crystallity capability of the constituent blocks, and also on the prehistory of the specimen [3,4].

Temperature transitions in polycarbonate –polytetramethylenoxide (PC PTMO) block copolymer resins, obtained in solutions by the method of acceptor – catalytic polycondensation are studied in the scheme:

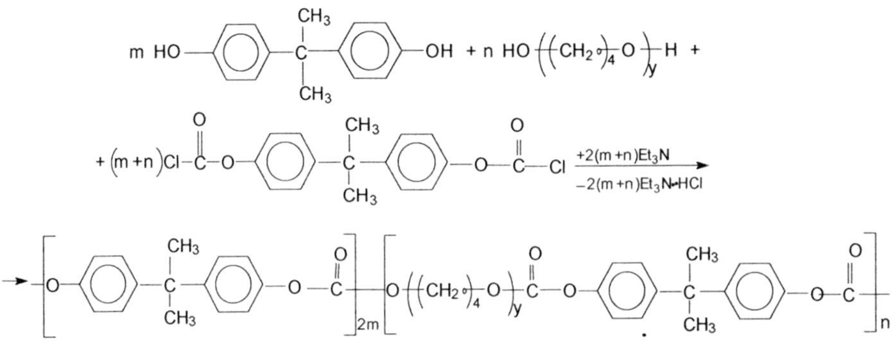

The method of differential scanning calorimetry (DSC) was used to define temperature transitions and evaluate the phase state of blocks in PC PTMO. Investigations of PC and PC PTMO block copolymer resins were carried out with the help of scanning calorimeter DSK-2 (by the firm "Perkin-Almer") in the temperature interval from 173 to 523 K at the heating speed 40° a minute. The measurement accuracy was 1-2 degrees. After the first scanning the specimen was quenched up to the primary temperature and the hardened specimen was scanned again.

The calorimetrical curves of PC and PTMO were taken for comparison. On the thermogram for PC there was only one change at 417 K, corresponding to the its vitrification temperature. The transition, connected with PC melting was absent, i.e. polycarbonate film had a liquid –like structure. However, as it has already been pointed out, according to the literature data, PC crystallizes in certain conditions and its melting temperature varies in the interval from 493 to 503 K [2].

For PTMO with the molecular weight 2000 and terminal hydroxyl groups the following temperature transitions are found: T_v=192 K, T_m= 300 K, Q_m= 21,4 cal./g.

To analyze the data, obtained with the help pf DSK for PC PTMO block copolymer resins, it was synthesized a model polymer, consisting of PTMO blocks with MM=2000, elongated with the help of bischloralformate bisphenol A. It turned out that temperature transitions of this model differ from transitions in PTMO with hydroxyl groups: T_m=294 K, Q_m=23,2 cal./g., $T_{cryst.}$= 268 K.

Polycarbonate –polytetramethylenoxide block copolymer resins were investigated in the form of films, obtained by the method of sprinkling from the solution, and in the form of powders. As the specimens had different prehistory, all of them were brought into similar conditions by heating up to 57 K and quenching (320 degr./min) to 153 K.

Calorimetrical investigations showed that PC and PTMO blocks in such block copolymer resins are partially combined. The dependences of temperature transitions are shown in Table.

Table. The main temperature transitions in PC PTMO block copolymer resins

Structure of PK-PTMO block copolymers, % mass.	T_g PK phase, K	T_g PTMO phase, K	T_m PK phase, K	T_m PTMO phase, K
100:0	422	-	493	-
90:10	415	-	-	294
80:20	403	-	-	293
70:30	395	-	-	295
60:40	343	189	493	291
50:50	338	191	494	290
40:60	323	195	493	289
0:100	-	192	-	294

As it was shown, the mean temperatures of vitrification, which are different form T_v of homopolymers, are observed for all the specimens in a high temperature area. With the increase of content of PTMO blocks the meanings of T_v move to the area of a low temperature. Moreover in BCR, containing up to 40% PTMO blocks, a partial combination of PC and PTMO blocks take place and T_{v1} of plasticized PC phase and T_{m2} PTMO phase is observed for them only. If contents of PTMO blocks in BCR is more than 30% of its weight, a more complex apportionment into separate phases, which are characterized by their vitrification temperatures (T_{v2}) and melting (T_{m2}), is observed.

Simultaneously, the melting temperature of crystallite PC phase (T_{m1}) develops. Temperature T_{v1} is still lowering and this fact testifies the partial compatibility of PC and PTMO blocks and in the area of BCR constituents.

The melting degree of the crystallity phase formed of PC blocks, and the degree of crystallity rise with the increase of contents of PTMO blocks. The maximal degree of crystallity (about 16%) in the studied line is achieved under containing PTMO blocks at about 60% of its weight. It is found out that PC PTMO block copolymer resins are sensitive to the change of the regime heating-cooling. For instance, the repeated processes of heating-cooling lead to lowering of T_{v1} up to 5-7 degrees. We may suppose that the observed phenomenon is connected with the partial thermal – oxidative degradation of BCR, and with the reaction of interchain exchange. Both of them should lead to improving compatibility of PC and PTMO blocks. Under the repeated heating-cooling process of BCR, besides the lowering of T_{v1}, a lowering of the melting temperature is observed. The melting temperature of the PC phase does not change significantly.

The obtained results show that a microphase separation of hard and elastic blocks is observed in PC PTMO block copolymer resins. As PC and PTMO blocks can crystallize, both vitrification temperature and melting temperature of different phases is fixed.

REFERENCES

[1] Noshey, A. Mac-Grant, G. *Block-copolymers.* A walkover. Moscow, 1980. 478 p.

[2] Smirnova, O.V., Erofeeva, S.B. Polycarbonates. Moscow, *Chemistry*. 1975. 288 p.

[3] Kalashnikova, V.G. Peculiarities of structure formation in crystallizing block copolymers. *Scientific prognosis in the field of polymers*. Moscow, 1978.

[4] Aksenov, A.I., Aksenova G.S., Semenkina G.M and others. An article, published in *RG Chem*. 1981, 3p.

In: Polymers, Polymer Blends, Polymer Composites… ISBN 1-60021-168-2
Eds: A.K. Mikitaev, M.K. Ligidov et al., pp.155-158

Chapter 19

THE CALCULATION OF TEMPERATURE STRESSES IN POLYMERS

B. M. Yazyyev
The Strength of Materials Chair, RSBU 162,
Sotzialisticheskaya Street Rostov-on-Don

To study the kinetics of temperature stresses in polymers, to analyse the influence of various factors on the flow of the examined processes, and to model the relaxation behaviour in polymers, a nonlinear constitutive differential equation is used in the paper. This equation was proposed by G.I. Gurevich [1], who called it "the nonlinear generalized Maxwell equation" out of respect for J. Maxwell's ideas [2] that served as a partial basis for deducing the equation. Total deformation is regarded as the sum of elastic, viscoelastic and temperature deformations:

$$\varepsilon = \varepsilon_e + \varepsilon^* + \varepsilon_T \tag{1}$$

Hooke's law holds true for elastic deformations ε_e, temperature deformations ε_T are calculated using the formula

$$\varepsilon_T = \int_0^T \alpha(T)dT \tag{2}$$

while in a three-dimensional case the correlation of stresses, time and viscoelastic strains is described by the generalized Maxwell equation:

$$\frac{\partial(\varepsilon_{ik}^*)_s}{\partial t} = \left[\frac{3}{2}\left(\sigma_{ik} - p\delta_{ik}\right) - E_{\infty s}\cdot\left(\varepsilon_{ik}^*\right)_s\right]\cdot\frac{1}{\eta_s^*} \quad i,k = x,y,z\,;\ \delta_{ik} = \begin{cases} 0 & npu\ \ i = k \\ 1 & npu\ \ i \neq k \end{cases} \tag{3}$$

Here: the index s (s=1,2,3,...) indicates the viscoelastic strain spectrum constituent that corresponds to a certain relaxation time; p is the average stress; $E_{\infty s}$ stands for highly elastic strain modules; η_s^* is the relaxation viscosity defined by the formula:

$$\eta_s^* = \eta_{0s}^* \cdot \exp\left\{-\frac{1}{m_s^*}\left[\gamma_s^* p + \left|\frac{3}{2}(\sigma_{rr} - p) - E_{\infty s} \cdot (\varepsilon_{rr}^*)_s\right|\right]\right\}. \tag{4}$$

In this equation η_{0s}^* is the initial relaxation viscosity factor, proportional to the relaxation time, i. e. viscosity in absence of stresses; m_s^* is the speed module reflecting the influence of the strain rate on the speed of relaxation processes; γ_s^* is the volume factor that accounts for the influence of uniform expansion or compression on relaxation speed. The indices rr denote main strains and stresses.

Relations (3) and (4) are true for relaxing medium that has an elastic constant carcass, able to restore its unstrained state. For plastically (finitely) deformable medium in (3) and (4) $E_{\infty s} = 0$ is to be introduced.

All the parameters included in (3) and (4) are rather complex temperature functions, and it has been impossible so far to determine these dependencies theoretically. The dependencies under question as well as the parameters themselves are obtained from macroexperiments on isothermal uniaxial tension, compression or shift in various modes at various temperatures.

Relying on the analysis of the experimental data, a number of works [3,4 et al.] prove that for the description of long-standing relaxation processes in polymers under quasi-static tests two constituents of viscoelastic strain are to be taken into consideration in (3) and (4) — a fast 'older' ε_1^* constituent and a slow 'younger' ε_2^* component. As a rule, for the description of relatively short-term quasi-static processes merely one - the 'older' - constituent is sufficient; its relaxation time is significantly shorter than for ε_2^*.

If in (4) the exponent is formally substituted by 'one', we get a linear (or linearized) equation of connection that for one member of the spectrum can be recast in the following form:

$$\frac{\partial \varepsilon_{ik}^*}{\partial t} = \left[\frac{3}{2}(\sigma_{ik} - p\delta_{ik}) - E_\infty \cdot \varepsilon_{ik}^*\right] \cdot \frac{1}{\eta_0^*} \tag{5}$$

As it has been noted, for elastic materials (for example, certain metals) $E_\infty = 0$ is to be introduced in (5). Then, in a one-dimensional case, the well-known linear Maxwell equation is achieved:

$$\frac{\partial \varepsilon}{\partial t} = \frac{\sigma}{\eta_0}$$

On the grounds of the above-mentioned equation of connection (3) we subsequently provide the solution of the problem that concerns the determining of temperature stresses in a polymer rod at temperature changes according to the linear law:

$$T(t) = T_0 + kt \,.$$

While solving the problem, the following assumptions are made: the temperature field in the rod is homogenous; the ends of the rod are rigidly fixed while in the rod the uniaxial stress state is observed ($\sigma_x \neq 0$); the plane section hypothesis holds true; volume strains (or the average stress) exhibit an insignificant influence on the relaxation process speed; two members of the relaxation times spectrum are taken into consideration.

As a result of these assumptions for the problem under question, equation (3) in consideration of (4) takes on form:

$$\frac{\partial \varepsilon^*_{x,s}}{\partial t} = \frac{\sigma_x - E_{\infty,s} \cdot \varepsilon^*_{x,s}}{\eta^*_{0,s}} \exp\left|\frac{\sigma_x - E_{\infty,s} \cdot \varepsilon^*_{x,s}}{m^*_s}\right| \tag{6}$$

For the total deformation increment in accordance with (1) there is:

$$d\varepsilon_x = \frac{d\sigma_x}{E} + \sum_{s=1}^{2} d\varepsilon^*_{x,s} + \int_0^T \alpha(T) \cdot dT \tag{7}$$

The equilibrium equation and the Cauchy formula look like:

$$\frac{\partial \sigma_x}{\partial x} = 0 \;;\; \varepsilon_x = \frac{\partial u}{\partial x} \tag{8}$$

Thus, for the unknown σ_x , ε_x , $\varepsilon^*_{x,s}$, u a complete system of equations has been derived. The system describes the history of the rod stress and strain conditions.

Assuming that before heating (cooling) the rod was in the equilibrium (limp) state, the following initial conditions can be written as:

$$t = 0 \;;\; T = T_0 \;;\; \sigma_x = 0 \;;\; \varepsilon^*_{x,s} = 0 \;\; (s = 1,2) \,. \tag{9}$$

The conditions at the rod edges will look like:

$$x = 0, l \;;\; u = 0 \tag{10}$$

$\varepsilon_x = 0$ is derived from the second equation (8), or in the differential form:

$$d\varepsilon_x = 0 \tag{11}$$

System (6) - (8) under the initial conditions or under condition (11) is solved using the numerical method of 'layer' integration enunciated in [5] and applied in many subsequent works [6 et al.]. The calculation is performed at each time 'layer', if the time increment Δt is constant. If Δt is a variable, then $t_i = t_{i-1} + \Delta t_i$ $(i = 0,1,2,...)$. For i -layer at $\Delta t = const$ a system of algebraic equations is obtained:

$$\left.\begin{array}{c} \Delta T = k \cdot \Delta t \\ \Delta\sigma_i = -E_i\left(\Delta\varepsilon_i^* + \alpha_i \cdot \Delta T\right) \\ \Delta\varepsilon_i^* = \sum_{s=1}^{2} \Delta\varepsilon_{s,i}^* = \sum_{s=1}^{2} \left(\frac{d\varepsilon_s^*}{dt}\right)_{i-1} \cdot \Delta t \\ \sigma_i = \sigma_{i-1} + \Delta\sigma_i; \quad \varepsilon_i^* = \varepsilon_{i-1}^* + \Delta\varepsilon_i^* \\ \left(\frac{d\varepsilon_s^*}{dt}\right)_i = \frac{\sigma_i - E_{\infty s,i} \cdot \varepsilon_{s,i}^*}{\eta_{0s,i}^*} \exp\left|\frac{\sigma_i - E_{\infty s,i} \cdot \varepsilon_{s,i}^*}{m_{s,i}^*}\right| \end{array}\right\} \tag{12}$$

These equations are to be supplemented with the initial conditions:

$$i = 0 \ ; \ t_0 = 0 \ ; \ T = T_0 \ ; \ \sigma_0 = 0 \ ; \ \varepsilon_{s,0}^* = 0 \tag{13}$$

REFERENCES

[1] Gurevich, G.I. 1974 Media Deformability and Seismic Wave Transmission. Moscow, *Nauka*. p. 483.

[2] Maxwell, J.Cl. 1927 *The Sci. papers of J.Cl.* Maxwell. Paris, Hermannp. p. 616.

[3] Rabinovich, A.L. 1970 The Introduction into the mechanics of Reinforced Polymers. Moscow, *Nauka*. p. 482.

[4] Goldman, A.Y. *The Investigation of Deformation Mechanisms in Certain Rigid Network Polymers Bounding at Shifts*. Ph.D. Thesis, Moscow, 1966. p. 250.

[5] Andreyev, V.I. On the Rigidity of Polymer Rods at Creep. In: *The Mechanics of Polymers,* 1968, V.1, pp. 15-22.

[6] Turusov, R.A. *Mechanical Phenomena in Polymers and Composites* (in Formation Processes). Doctoral Thesis, Moscow, 1983. p. 363.

In: Polymers, Polymer Blends, Polymer Composites… ISBN 1-60021-168-2
Eds: A.K. Mikitaev, M.K. Ligidov et al., pp.159-162

Chapter 20

COMPOSITES ON THE BASIS OF POLYHYDROXIETHERS AND GRAPHITES

D. A. Beeva, A. K. Mikitaev, G. E. Zaikov, R. Z. Oshroeva, V. K. Koumykov and A. A. Beev*

[1]Kabardin-Balkarian State Academy of Agriculture, 360004, Nalchik, Tarchokov Street, 1a, Russia

[2]Kabardin-Balkarian State University, Chernyshevskogo Street, 173, Russia

ABSTRACT

By settle polycondensation the polymeric compositions were created, in which high crystalline graphite of the scaly form, using as filler was entered into polymer during synthesis of polyhydroxiethers.

The results of experiment showed, that in the presence of synthetic graphite the viscosity of polymer increases; this proves our earlier assumption, that selective adsorption leads to the increase of the local concentration of monomers on the surface of fillers and to the increase of polycondensation reaction speed.

Keywords: Polyhydroxiethers, compounds, polycondensation, graphite, heterophase conditions.

Polyhydroxiethers (PHE), representing high-molecular epoxy compounds, are perspective polymers because of a complex of valuable properties, first of all adhesion to various surfaces. They find their application as a basis of varnishes, glues, film-forming substances, constructional materials in electronic and electrical engineering industries, in automobile and ship constructing and in the number of other branches of industry.

---

*Correspondence to: 360030, KBR, Nalchik, st. Tarchokov, 1a, Kabardin-Balkar State Academy of Agriculture, faculty of chemistry, Beev A.A. mailto: beevaues@front.ru

However, the further expansion of areas of their application attributes to the creation of new, more perfect compounds with required operational characteristics such, as electric and heat conductivity at high indicators of physic mechanical and antifriction properties.

It was earlier informed [1] about the developed one-phase method of synthesis of PHE by settle polycondensation at heterophase conditions. The scheme of proceeding reaction at use of besphenol and epychlorhydrine as initial substances can be presented as follows:

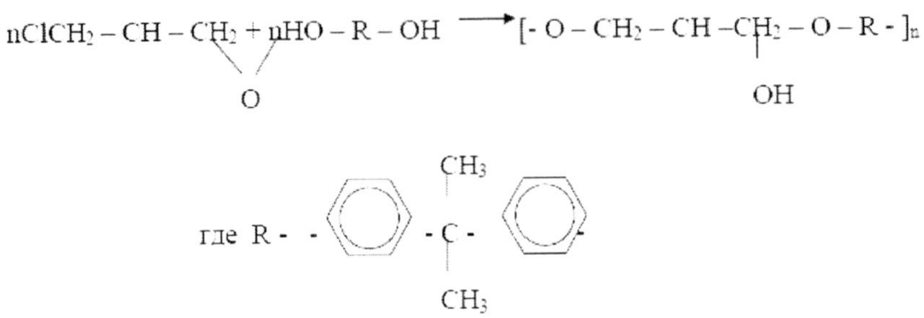

Polycondensation at the presence of solid inorganic compositions, on the one hand, carrying out a role of catalysts, and with another being fillers for formed polymers have both theoretical, and practical value and is connected to practical problems of production of polymeric composite materials.

Use of graphite is one of perspective directions of creation of composites on the basis of PHE. Absence of laws of formation of PHE-graphite boundary layers and regulation of adhesive interaction in system PHE - graphite, due to modifying of filler surface, defines a topicality of the present paper.

At present time methods of polymeric composite materials production, in which the filler is entered in polymer during synthesis, are widespread. Thus the filler influences not only on the kinetics of polymerizations or polycondensation, but also substantially defines the structure of polymer, and consequently, on the properties of a composite material.

By settle polycondensation the polymeric compositions were created, in which high crystalline graphite of the scaly form, using as filler was entered into polymer during synthesis of PHE.

In the presence of disperse particles with complicated surfaces, an adsorption of monomers on a surface of fillers takes place. As a result the local concentration of monomers at a surface of fillers increases.

It is known, that graphite has selective adsorption ability. Organic substances are adsorbed on a surface of graphite the more strongly, than more they contain aromatic rings. Therefore in presence of high crystalline graphite the selective adsorption of dioxydifenilpropan takes place.

At adsorption of besphenol on a surface of graphite the conjugation of π-electron clouds of aromatic rings of 4,4′ dioxydifenilpropan and graphite is possible. As a result of such interaction the nuclearphility of phenol hydroxyl group reduces and as consequence the reactionary ability of phenol hydroxyl groups also reduces.

On figure 1 the dependence of the reduced viscosity of PHE from the maintenance of graphite is illustrated. At the presence of graphite monotonous decrease in viscosity of formed polymer (a curve 1 on fig. 2) is observed.

With the purpose of check of the assumption about selective adsorption ability of crystal graphite, the synthesis of PHE in the presence of isotropic synthetic graphite of mark MPG-8 was processed. The results of experiment have shown, that at the presence of synthetic graphite there is an increase in viscosity of formed polymer (fig.1, curve 3). It confirms the assumption made earlier that as a result of selective adsorption the local concentration of monomers at the surface of feelers increases, that leads to the increase of speed of polycondensation reaction.

At oxidation of graphite, on its surface various, chemically active, oxygen containing groups, which irreversibly chemisorbs besphenol A, are formed. This is a result of chemical interaction of hydroxyl groups of besphenol A with surface groups of graphite and as consequence the decrease in the viscosity of polyhydroxiether synthesized at the presence of graphite (figure 1, curves 2 and 4).

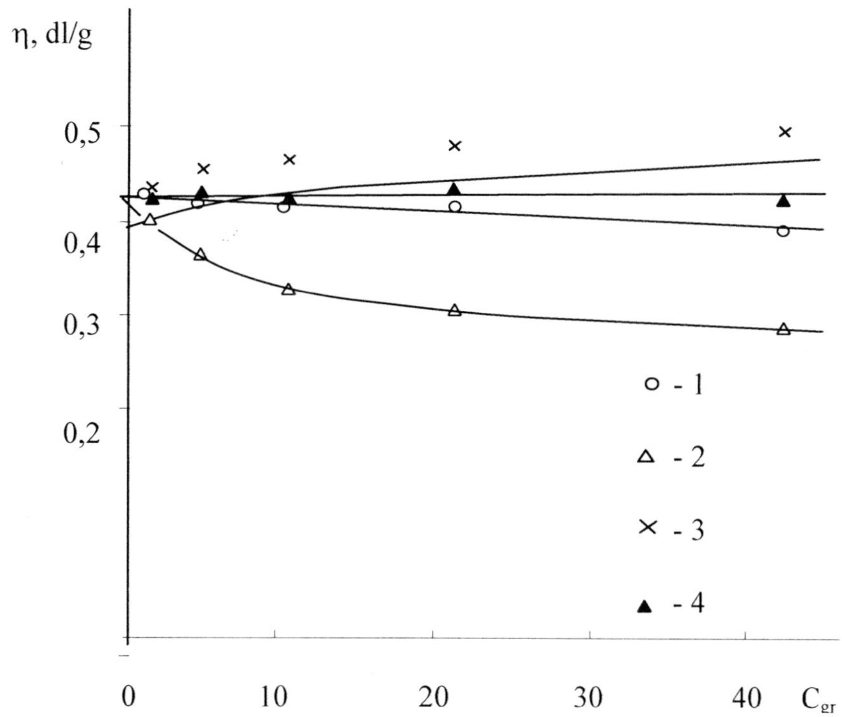

Fig. 1. Influence of graphite on the reduced viscosity of the obtained polymer at the presence of graphite. 1 - non processed GL graphite, 2 - processed GL graphite, 3 – non processed MPG graphite, 4 – processed MPG graphite

The indirect proof of influence of graphite processing on a polymeric matrix is the fact that at dissolution of a polymeric matrix of composites in chloroform there is a subsidence of particles of graphite in the samples made by mechanical mixture. For samples with the processed graphite, made by synthesis of polymer, steady enough suspension is formed.

Pressing at temperature 180°C of graphite, obtained after fivefold extraction of polyhydroxiether, has given the following result: all pressed samples of the raw graphite were scattered at extraction from the form, while oxidized, at least were not scattered at extraction from the form. It may testify the presence of polyhydroxiether chemisorption on the surface of the processed graphite.

Thus, the crystallity of graphite influences on the reactionary ability of besphenol A, and the big role is played the presence on a surface of feller, capable to interact with monomers, changing functionality of the last.

Thus, as a result of the lead researches the new method of obtaining of graphite feeling polyhydroxiethers with high parameters of adhesion to various surfaces, electrical conductivity, physic-mechanical and antifriction properties is offered.

The opportunity of improvement of physic mechanical properties of graphite feeling compositions due to reduction of aggregation of particles of feeler and improvements of adhesive interaction PHE and graphite as a result of graphite surface processing is shown.

REFERENCES

[1] Beeva, D.A., Beev, A.A., Mikitaev, A.K. *Synthesis of polyhydroxiethers on the base of besphenol // Modern scientific technologies.* –Moscow. -RAS-publisher. -2004. –P.80-81.

In: Polymers, Polymer Blends, Polymer Composites... ISBN 1-60021-168-2
Eds: A.K. Mikitaev, M.K. Ligidov et al., pp.163-165

Chapter 21

HEAT-CONDUCTING COMPOSITIONS ON THE BASE OF EPOXY POLYMERS

A. A. Beev[1]*, A. K. Mikitaev, R. Z. Oshroeva[2], D. A. Beeva[2] and V. K. Koumykov[1]

[1]Kabardin-Balkar State Academy of Agriculture, 360004, Nalchik, Tarchokov Street, 1a, Russia

[2]Kabardin-Balkar State University, Chernyshevskogo Street, 173, Russia

ABSTRACT

Heat conducting epoxy compositions on the base of phenol-formaldehyde new lacquer oligomers with low content of ion impurities and hydrolyzed chlorine are produced.

Keywords: Epoxy oligomer, electric insulation material, viscosity, heat conductivity.

Heat conducting epoxy compositions on the base of phenol-formaldehyde new lacquer oligomers with low content of ion impurities and hydrolyzed chlorine are developed.

Epoxy new lacquers were consolidating by corresponding new lacquer oligomers, mixing in relation 2:1 in presence of 1% 2-metilimidozol as consolidating catalyst. By yield of gel fraction the optimal step-by-step regime of consolidation was matched.

The results of experiment showed (fig.1 and 2, tab. 1), that the heat conductivity of epoxy new lacquer compositions may increase from 0,2 to 0,99 Wt/m·K. In this case the highest heat conductivity has the compositions filled by boron nitride. During the supplement to the system of aerosil in quantity of 2% the heat conductivity coefficients decrease; this fact may be explained by dilatation of structure and by formation of pores.

*Correspondence to: 360030, KBR, Nalchik, st. Tarchokov, 1a, Kabardin-Balkar State Academy of Agriculture, faculty of chemistry, Beev A.A. E-mail: beevaues@front.ru

Table 1. Heat conductivity of epoxy new lacquer filled compositions

BN, weight %	B_4C_3, weight %	λ, Wt/m·K
5	25	0,49
10	20	0,58
15	15	0,60
20	10	0,67
25	5	0,99

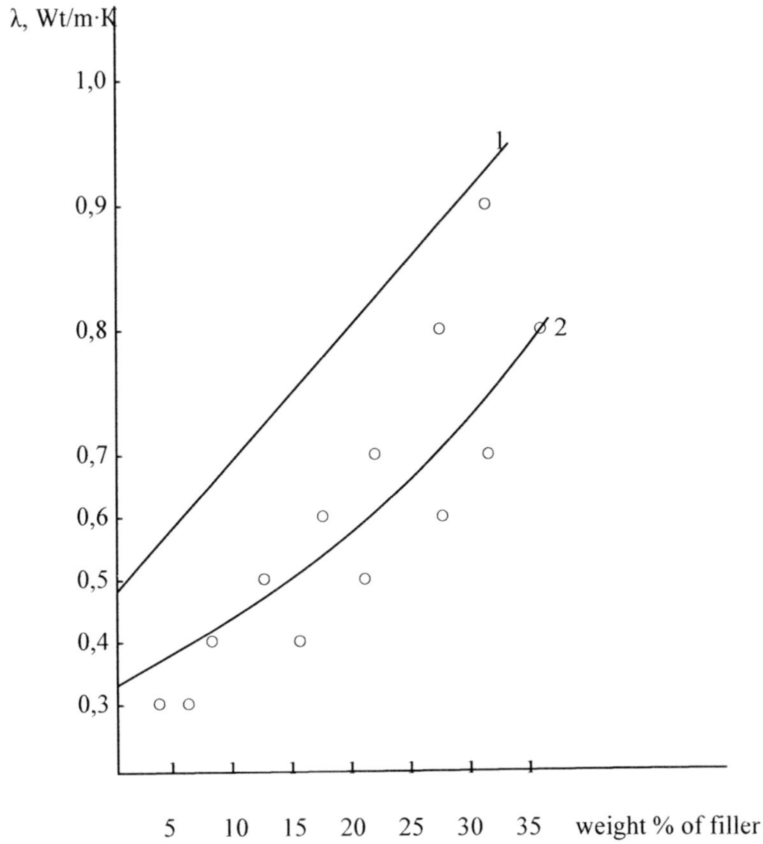

Fig. 1. The dependence of heat conductivity of compositions upon the content of boron nitride (1) and effect of aerosil additives (2)

The heat conductivity of epoxy compositions, filled by boron carbide, lower than those filled by boron nitride (tab. 1).

As it is shown on the figure 2, the compositions with filler particles with sizes from 20 to 150 micrometers more heat conductible than those with the sizes from 150 to 250 micrometers.

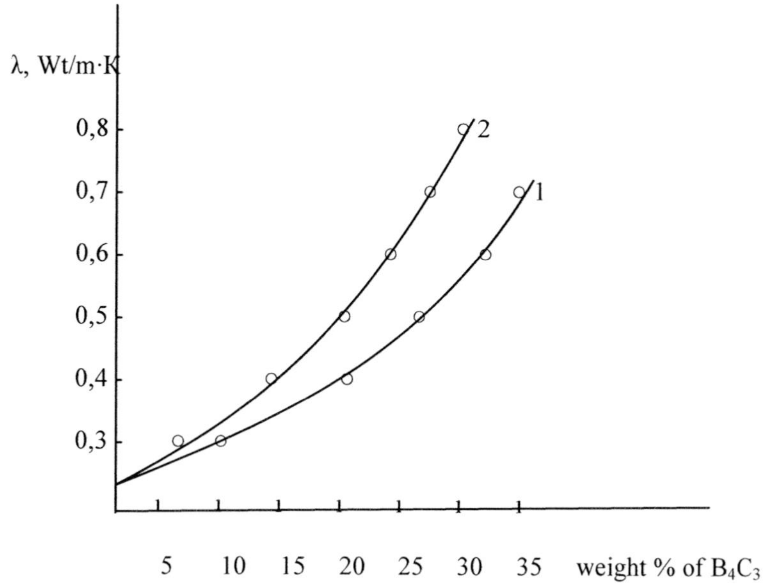

Fig. 2. The dependence of heat conductivity of epoxy new lacquer compositions upon the sizes of filler particles. 1 – (150-250) mcm; 2 – (20-150) mcm

Heat-resistance and dielectric properties of the compositions with highest heat conductivity were studied. The thermo gravimetric studies showed that samples without fillers decompose at 340-345^0 C and samples with fillers – at 320-325^0 C.

During determination of dielectric constant (E = 3,5-5,0), tangent of angle of dielectric loss (tgδ = 0,02–0,03), specific volume electrical resistance (ρ = 10^{-15} – 10^{-17} Om·cm) we came to the conclusion that nitride and carbide of boron do not deteriorate the electrical properties of compositions and they are useful for pressurization of semiconductor instruments of high reliability.

In: Polymers, Polymer Blends, Polymer Composites... ISBN 1-60021-168-2
Eds: A.K. Mikitaev, M.K. Ligidov et al., pp.167-169

Chapter 22

FILLED LOW VISCOSIVE EPOXY COMPOSITION MATERIALS

A. A. Beev[1*], A. K. Mikitaev [2], R. Z. Oshroeva[2], V. K. Koumykov [1] and D. A. Beeva[2]

[1]Kabardin-Balkar State Academy of Agriculture, 360004, Nalchik, Tarchokov Street, 1a, Russia

[2]Kabardin-Balkar State University, Chernyshevskogo Street, 173, Russia

ABSTRACT

On the base of accessible epoxy-dian oligomer ED-20, fillers and other ingredients high technology fill up compounds with valuable complex of exploitation properties are obtained.

Keywords: Epoxy oligomer, electric insulation material, viscosity, heat conductivity.

Low heat conductivity of polymer electric insulation materials limits the field of their use, especially for the purpose of pressurization of radio electronic instruments. That's why the development of high heat conductive filling epoxy compounds was done.

For filling compounds producing one of the most accessible epoxy dian oligomers ED-20 was chosen as an initial reagent. It has 20-21% epoxy groups contaminant.

Among the requirements to the base of composition, providing its high effectiveness, are: low viscosity, run out ability through the nozzle of 0,5 mm diameter. At the same time the base has to have the viscosity, which eliminates the filler subsiding. The life time of composition has not to be less than 8 hours and the composition has to be soluble in water.

*Correspondence to: 360030, KBR, Nalchik, st. Tarchokov, 1a, Kabardin-Balkar State Academy of Agriculture, faculty of chemistry, Beev A.A. E-mail: beevaues@front.ru

Proceeding from above mentioned requirements the optimal formula of composition was matched. Among known epoxy oligomer hardeners the most satisfied to the set problem is the isometiltetrahydroftal anhydride, because of its liquid consistence.

After matching of hardener system, compositions, containing the mixture of above mentioned hardeners with epoxy oligomer, were prepared. Viscometric measurements showed, that despite the fact that the composition is a mobile liquid (η = 493,6 Cst) it still has high viscosity. For viscosity reduction a styrol was matched as a diluent.

On fig. 1 the curve of dependence of composition viscosity upon styrol concentration is shown.

The influence of fillers on the heat conductivity of epoxy compositions was studied. The heat conductivity of pure composition without filler was determined; it ranges from 0,2 to 0,25 Wt/m·K in wide temperature interval.

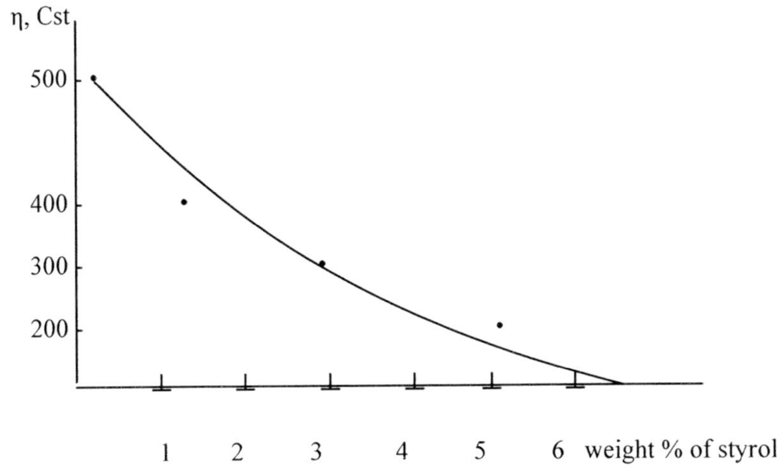

Fig. 1. The dependence of composition viscosity upon styrol concentration

On fig. 2 the dependence of heat conductivity of epoxy compositions on filler content is shown.

The increase of filler content courses the increase of heat conductivity; at the same time the viscosity of composition is also increases, which makes difficult its processing.

During the studies, which were done in wide temperature interval from 0^0C to 200^0C, the maximums on the curves of temperature dependence of heat conductivity were detected. These maximums correspond to the glass-transition temperatures.

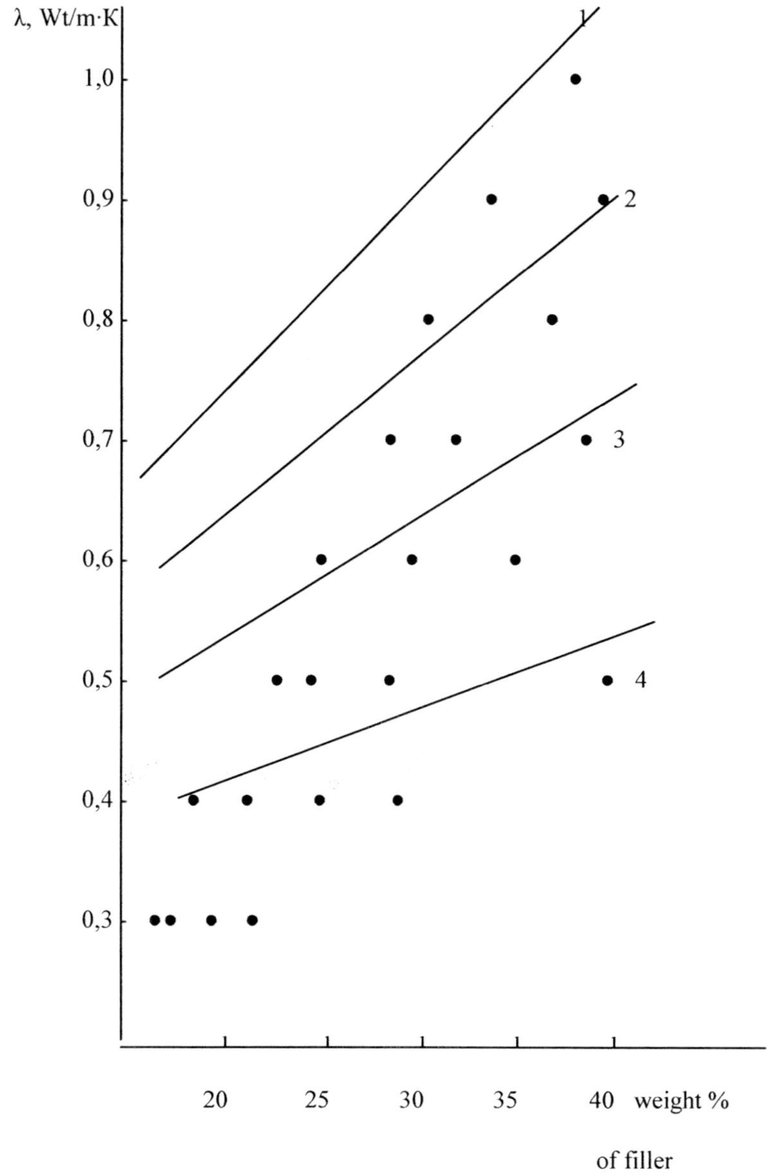

Fig. 2. The dependence of heat conductivity of epoxy compositions on filler content. 1-SiC; 2-33,3 % BN + 66,7 % Al_2O_3; 3-50% SiO_2 + 50 % Al_2O_3; 4-SiO_2

It was established that for obtaining epoxy compositions with maximum heat conductivity it is necessary to keep samples in low vacuum at pressures 50-100 mm.mer. during 15-20 min.

Otherwise the heat conductivity of samples is low and its concentration dependence is extreme. This may be explained by the fact that high loading of filler leads to bubble formation in casting.

The studies of dielectric properties showed, that the developed compositions maintain its properties in the temperature interval from 20^0 C to 150^0 C. Dielectric constant of compositions is 4-5 tangent of angle of dielectric loss - 10^{-2}, specific volume electrical resistance - 10^{-15} Om·cm. These data corroborate, that heat conductive compositions are useful for pressurization of electronic instruments.

In: Polymers, Polymer Blends, Polymer Composites... ISBN 1-60021-168-2
Eds: A.K. Mikitaev, M.K. Ligidov et al., pp.171-175

Chapter 23

THE ELECTRICAL CONDUCTIVE COMPOSITIONAL MATERIAL WITH LOW INFLAM ON POLIPROPILEN BASIS

G. M. Danilova-Volkovskaya* and E. H. Amineva
The Rostov-on-Don state Academy of Agricultural Mechanical Engineering,
1, St. Soviet Land, Rostov-on-Don, 344023.

ABSTRACT

Efficiency of application of the method based on receptions of optimum planning of experiment for reception of composite materials on the basis of polypropylene with the set complex of properties is proved. Studying influence carbon наполнителей on properties of electrowire composite materials is lead.

Key words: the method based, polymer compositional materials, electrical conductive, polypropylene, technical carbon, acethylene soot.

Today it is important the problem of electrical conductive polymer compositional materials with low inflame elaboration. The main direction of using these composite materials is to manufacture reining electrodes using in aggressive environment and law temperature heating appliances.

Because it make to provoke these heating and kindling. All over the worlds it is shown the perspectives of receiving electrical conductive inflammable polypropylene materials.

To make electrical conductive polymer material it is necessary to use special types technical carbon. Unlike usual soot it has more specific surface. Industrially producing plumbago has the size 100-1000 times lager then acetylene soot. In fact, to achieve necessary level of electrical conductivity we need use 10-15 % mass of plumbago AG-4 in material 15-20% of soot.

---

* E-mail: *danilova-volk@yandex.ru* The state Novorossiysk marine academy, 93, pr. Lenin, Novorossiysk, 353900.

It was investigated the influence of the given carbon on characteristics of electrical conductive materials with low inflam on base of polypropylene (PP).

This investigations showed us that in these interval it filling concentrations (up to 50% mass) the carbon fillings ensure electrical conductivity like metal powder (diagram 1).

Soot carbonfilled PP has the most electrical conductivity because the soot particles involve a lot of polymer in a state of transitional stratum and increase the concentration plumbago in unsystematic spheres. These regularities are inspected by changing thermophisical properties. Energetic increasing of thermal conductivity PP filled with plumbago 50% mass and more is consequence of contract between the filling particles. The most effect of thermal conductivity is seen by introducing plumbago in material containing 15-20% mass of soot.

Measuring the thermo coefficient of electrical resistance PP showed for these materials it is positive quantity.

Investigation the fotoes which was got by electronic microscope has showed that in case of small capacity of polymer (5-10%) the filling is distributed under the pretext of separate particles or small anglomerates. Here we can see transitional layer of PP. The latter has orientated structure and width of 20-25 mm.

Investigation the mechanism of conductivity of composite material confirm us that on the border of contact between soot and PP the double electric layer is formed which depended upon injection of charge bearer from soot into polymer. In case the diffusion length of injected charge in polymer layer is more than a half distance the soot particles we can see existence of electricity on condition that electric tension. It is necessary to insure steady distribution soot in PP and to decrease quantity at agglomerate of filling to realize injective conductivity in polymer through thin layers of polymer between the particles of filling. Including modifying additions at the beginning and at extrusion leads to good mixing.

Choiceing the modifying addition for electrical conductive unflam composite material it is necessary to take into concideration the effectivity of its influence into PP characteristics (into changing of electrical conductivity. Preliminary trials showed the best results can be achieved in case of introduction of silicon of organic rubber (SKTN-A).

By passing of electrical currents the composite may be heated. To avoied kindling it is necessary to choice the untipierens.

The mechanisms of putting out the flame reactions by galogen supported combinations were investigated.

The bad combination of flame moderates of PP provides their intensive migration on surface. It leads to reduction of effectivity of fire-resistance. It is known that reduction capacity of polymer matrix and creation more firm structure of PP decelerate the process of untipiren migration.

It is known that in silicon organic fillings presence the stability of composites to thermo-oxidise destruction is increased. We have an information that galogen compositions as ingibitors of fireing of PP are quiet effective.

Among them decabromdifeniloxid was chosen. Threeoxid of surema was chosen as a synergist galogenconsisted antipirens.

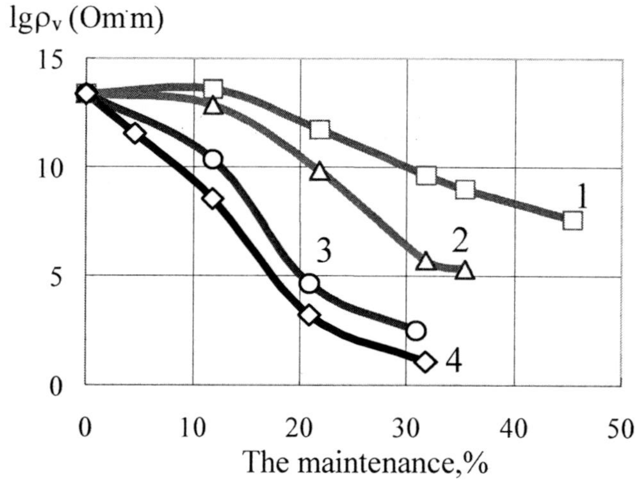

Fig. 1. Dependence of specific volumetric electric resistance of samples of composites on the maintenance: 1 - a powder of copper, 2 – graphite AG-4; 3 - acethylene soot; 4 - acethylene soot +graphite AG-4

It was investigated the opportunity of using disc extrusion for getting things from composite material on modifuing PP bases having electrical conductivity and low inflame.

The choice of process of disc extrusion is depended upon the opportunity of realization of high deformation. The optimisation of composition of multycomponent electrical conductive composite material on modified PP bases with low inflame, getting adequate regressive models were held by methods of optimal planning of experiment.

A lot of factors cause difficulties in carrying experiments for optimization of technological modes and prescription of compositional materials: too expensive raw materials and equipment, etc. That is why it is necessary to elaborate new method of optimization of recipe of composites.

The most actual and perspective task to solve this problem is using methods of mathematic statistics, exactly the planning of optimal experiment. The advantage of thematic models describing the influence the consistence of composite material into its quality in case of simultaneous varying of concentration of several components. These experimental investigations give us opportunity to get identical physical mathematic models "consistence quality" without carrying full set experiments.

These models give us opportunity to have scientific optimization of recipe of multycomponent composite materials and conditions of its conversion into things with specific quality.

Such planning of the concentration of this components of composite materials on PP bases was carried on bases of investigation of influence the parameters of different types of fillings into quality of modified PP.

The presents of acetylene soot- Xi, threoxid surema - Xi, decabrom-feniloxid - - X_3 were the independent variables. The response function is quantity of measureing characteristic in the time of verifying independent variables.

The response function were strength in rapture -Yi, comparative lengthening -¥2, destructive effort during bending -Y_3, specific work of percussion stickiness -Y4, PTR - Ys, specific electric resistance -Ye, oxygen index -Y_7

According to experiment planning the samples of compositional materials were got and their qualities were investigated. It got us an opportunity to determine the meanings of coefficients of regressive equation. The model of linear square equation was created according to intercourse of changeablequantity. This model chows the changer of qualities of compositional material according to its constitution:

$$\begin{cases} Y_1=10{,}05-1{,}73X_1+2{,}36X^2_1+0{,}87X_2-1{,}34X^2_2-0{,}54X_3+0{,}73X_1X_3; \\ Y_2=5{,}97-1{,}22X_1+1{,}35X^2_1-3{,}03X_2-1{,}03X_3; \\ Y_3=13{,}67-1{,}56X_1+6{,}63X^2_1+1{,}99X_2+4{,}23X^2_2+3{,}38X_3-0{,}66X_1X_2-1{,}63X_2X_3; \\ Y_4=6{,}66-1{,}05X_1-0{,}83X^2_1-0{,}32X_2-0{,}54X_3-1{,}33X^2_3+0{,}87X_1X_2; \\ Y_5=10{,}42-0{,}19X_1+0{,}51X^2_1-0{,}26X_2-0{,}24X_3+0{,}71X_2X_3; \\ Y_6=9{,}57+1{,}49X_1+0{,}42X_2+0{,}25X^2_2-0{,}79X_1X_2; \\ Y_7=10{,}67-0{,}53X_1-0{,}01X^2_1-0{,}47X_2-0{,}38X_3+1{,}56X_1X_3. \end{cases}$$

Significance of coefficients of the system was held according to the "Students criterion". The valuing of identicality of regressive equations was held according to the "Fisher's criterion" using valuing of probability 0,98.

At the end of our investigation it is necessary to solve the task of multy-criterium optimization of material consistence. Harryngton's function of de-sirebility was held as a general criterion.

The control experiment was held to prove calculated facts. This experiment determined the similarity of results and using the elaborated math-ematic model.

As a result the recipe of compositional electrical conductive material with low inflame on modified PP bases was received.

The calculated and received facts differed nomore than 5% from each other. It proves the optimum of calculated recipe and optimisational method. The noticed recipe of compositional material differs from traditional. The level of conductivity we need are ensured by consistens of electrical conductive fillings 35%. Its deformational quality and strength do not change for worse but increased.

The recipes of compositional materials on modified PP bases was alabo-rated with the help of this method: dispersefilled PP with improved technological qualities, electrical conductive materials, electrical conductive materials with increased frost resistance and improved pressing characteristics.

That is why the effectivity of using of this method is proved because it is based upon optimal planning of experiment to get compositional materials on PP bases with the complex of properties we need.

Reference

[1] Danilova-Volkovskaya, G.M. Electrospending material with the lowered combustibility on the basis of polypropylene. *"Plasticheskiye massy"*.—2002.— № 3. — C.41— 43.

[2] Danilova-Volkovskaya, G.M. Heat-physical properties of polymeric composite materials: the Manual. *PXTУ.* Moscow 2003. — 30 c.

In: Polymers, Polymer Blends, Polymer Composites… ISBN 1-60021-168-2
Eds: A.K. Mikitaev, M.K. Ligidov et al., pp.177-182

Chapter 24

RESEARCH OF MIXES ON THE BASIS OF CORN STARCH AND POLYETHYLENE

Madina L. Sherieva*, Gennadi B. Shustov, Ruslan A. Shetov, Betal Z. Beshtoev and Inna K. Kanametova
Kabardino-Balkarian State University, Nalchik

ABSTRACT

The aim of the research was to get a number of composition based on corn starch and a synthetic polymer – polyethylene and further study of physical and mechanic properties behaviour under the influence of aggressive mediums and biodecomposition in the soil as well.

Keywords: starch, polyethylene, composition, physical and mechanic properties.

Now, when the rates of growth of plastics manufacturing are extremely high, it is especially important, that development of the plastic industry take into consideration the problem of use of the plastics which are not outdated yet, but have lost their initial properties or consumer value of products, such as the thrown out products, and also technological waste products of their manufactures [1]. The problem under consideration is acute and relevant, it has ecological and economic aspects as it is connected with environmental protection and rational use of natural resources, needs for reduction the price of raw material for manufacture of polymers. At the same time it is the least investigated [2].

The aim of our researches is to get a number of compositions on the basis of corn starch and polythene.

While preparing the compositions on the basis of PEHD (M-273) and corn starch (GOST 12020-72) there were received compositions with the maintenance of starch from 1,5 % up to 15 % [3]. The mixture of polythene and starch, preliminary plasticized with a quantity of

*Kabardino-Balkarian state university, Nalchik 360004 Nalchik st. Chernyshevskaya, 173. sheri12@rambler.ru

glycerin, is loaded into the plodder; the temperature in the auger cylinder being 190 o C, the given mixture is fused and then the received melt is divided into grains. It is worth mentioning, that after the extrusion of the polythene and starch composition, the extrudate turns out to have thin foam-like structure. Therefore the received samples were under the same conditions once again.

Then some characteristics were investigated. For this purpose from the extruded composition and the initial polythene using the method of pressing test samples which represent transparent films were made. It is possible to judge the character of the occurred changes by the results represented in tables 1-4 and on figures 1-3.

Researches of the IR-spectra films of the initial polythene and the received compositions reveal that in the process of extruding there take place some changes in the field of 1300-900 sm - 1. In process gelatinization and destroying of starch when extruding compositions there occur changes of dielectric properties. Besides the composition structure, properties of film samples are also influenced by frequency rate of extrusion's. Thus introducing of starch raises polarity and values of a tangent of a corner of dielectric losses.

Introducing of starch and frequency rate of extruding influence the parameter of a melt's fluidity and explosive durability differently (tab. 1).

Table 1. Physic-mechanical properties of the pressed samples of compositions on the basis of polythene and starch

	Structure of composition, %		MFR $^{190}_{21,6}$	σ_{pi}, МПа	ε_{sp}, %at tearing up
	polyethylene	starch			
1	100	0	6,36	36,3	>500
2	98,5	1,5	17,57	17,7	35
3	97	3	34,87	17,7	53
4	95	5	45,93	17,7	27
5	93	7	37,94	15,1	15
6	90	10	31,5	10,8	9
7	85	15	17,06	16,7	12

The researches made in the sphere of creating and manufacturing biodecomposed polymers, are not only of theoretical, but also of applied character. So, within the framework of the given work, there was used the corn starch made at the open joint-stock company " KSF " (KBR Maiskiy region, village Aleksandrovskaya). Researches of the electric properties of the samples received by pressing are reflected on fig. 1-3.

It is seen on the diagram that values of tgδ are constant up to 120 °C, values of tgδ taking into account this frequency 104Гц (10-3-10-2) correspond with those given in literature. This correspondence is important from that point of view that then the observations and conclusions concerning the composition PE+S are possible to be applied to a great extent to other polyolefins.

At temperature higher than 120 degrees a rise of dependence of tgδ on T with a possible peak at 190 °C is observed. The specified temperature dependence essentially changes when introducing starch (fig. 2). For example, at its maintenance in 1,5 % background values raise a little. The background area extends. The planned peak at temperature of 190 °C disappears, but the precise maximum is found out at 85-90°C. As this peak didn't take place for the initial

PE, it can be related either to starch, or to the properties of the composition PE+S proper. This assumption is proved at considering of the diagram tgδ on T composition PE+3 %S. There are already 2 low-temperature peaks: approximately at 45 °C and 100 oC. These observations allow to assume the intensification of the influence of the additives on properties of the composition at these concentration already.

This intensification of the contribution of starch to the properties of the composition is seen in some way when studying structures with higher maintenance of starch.

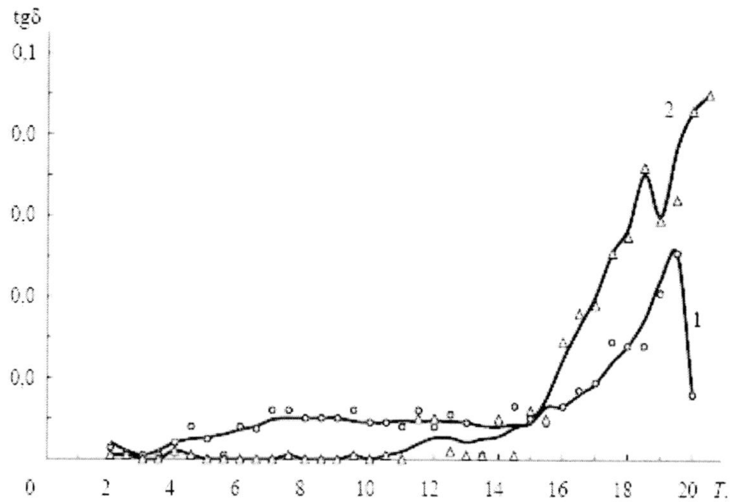

Fig. 1. Dependence of a tangent of an angle of dielectric losses tgδ on temperature T for granulated samples of initial not stabilized PEHD (M-273). Modes of preliminary heat treatment: $T = 100^{0}$C vacuum, 5 hours (1) and $T = 100^{0}$C without vacuum, 1 hour. (2). Frequency - 10 кГц

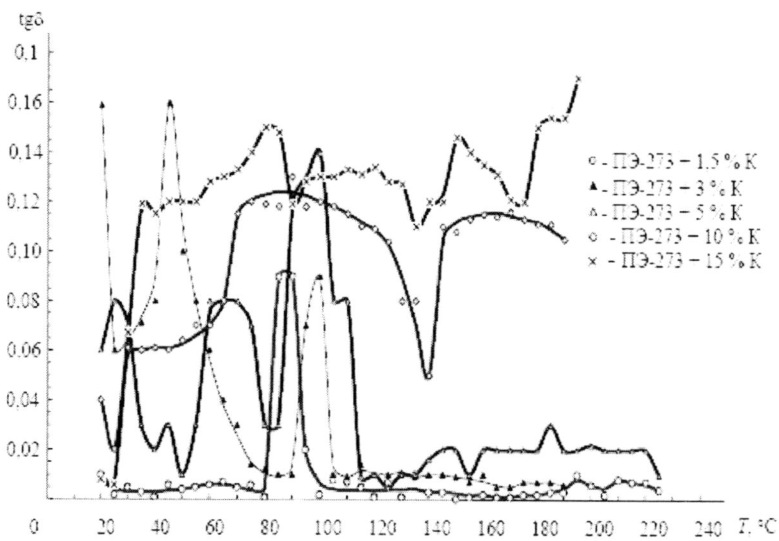

Fig. 2. Dependence of a an angle of dielectric losses tgδ on temperature T for compositions PEHD (M-273) + starch. Frequency - 10 кГц

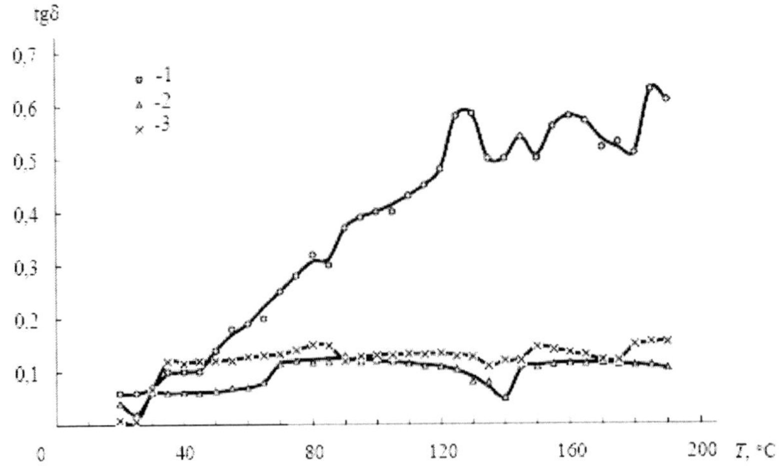

Fig. 3. Dependence of a tangent of an angle of dielectric losses tgδ on temperature *T* for compositions PEHD (M-273) + starch. The maintenance of starch - 7 % (1); 10 % (2); 15 % (3). Frequency - 10 кГц

Thus, at its 5 % maintenance several (3-4) low-temperature peaks already reveal, besides that the general background of tgδ values increases and on all the temperature interval tgδ is not less, than 10-2 (fig. 2).

At 10 % the maintenance of starch the background of tgδ values suddenly increases (5-50 times in comparison with 5 %). Low-temperature peaks merge into one wide (25-130 °C) peak. It is obvious, that this structure, and in the even greater degree the structure with 15 % maintenance of starch on dependence tgδ on T reveals properties of friable structure, possibly polar to the maximum and easily destroyed in the long term.

It is worth considering separately the compositions with high maintenance of starch in comparison. On fig. 3 the dependence of tgδ on T for compositions with the maintenance of starch 7, 10, 15 % is resulted. It was revealed rather unexpectedly, that referring the structure with 7 % of starch even on the background of tgδ values from 0,05 up to 0,15 (10-15 %K) dielectric losses of composition PE+S are very high in all the temperature interval, beginning from 35 oC and is higher.

Obviously, it is the most "bad" composition, remembering the destructive effect of starch on the initial PE. The question which is now put before us as researchers, consists in the following. It is necessary for us to create a composition which would be easily biodecomposable to the maximum, but at the same time the same composition should keep its properties during the necessary term. Thus, it is necessary to pick up an optimum structure of such composition, but for this purpose researches of the maximal set of physical and chemical properties of the composition PE+S are required.

The effect of aggressive mediums on the received samples in accordance with GOST 12020 were investigated as well. As aggressive mediums were used:

HCl - 10 % solution
NaOH-10 % solution
H2O - distilled water

It is known, that polythene is inert at action of many chemical reagents, namely, does not react with alkalis of any concentration, with solutions of neutral, sour and basic salts, organic acids (for example, with ant or acetic), with solutions of salts - oxidizers (for example, potassium permanganate) and even with the concentrated hydrochloric and fluoric acids [5].

Hence, the increase in weight of samples when keeping in solutions of 10 % hydrochloric acid and 10 % sodium hydrate solutions is caused by hydrolysis of starch in the beginning up to dextrins, and at full hydrolysis - up to D-glucose [4]. When keeping the samples in water grains of starch, contained in the composition, collapse with formation of paste, then swell, attaching small amounts of water (it is a convertible stage) [5]. It is proved by gradual increase of samples in weight at immersing into water for 3-18 days.

BIODECOMPOSITION IN GROUND

Biodecomposition in ground was defined at keeping the received pressed samples in ground (pH=6,88, the maintenance of humus - 0,16 %, exchange acidity =25,87 mg.экв./ in 100 gr. of ground) during 48 day. Then the study of their rheological and deformation - strengthening characteristics was carried out. The results are given in tab. 2-4.

Table 2. Change of explosive pressure of the pressed samples of compositions on the basis of polythene and starch at biodecomposition

№	Structure of composition, %		σ_p, МПа,	σ_p, МПа,	σ_p, МПа,	σ_p, МПа
	polyethylene	starch	исх.	in 14 day.	in 28 day.	in 42 day.
1	100	0	36,3	35,8	36,2	35,9
2	98,5	1,5	17,7	18,1	19,4	19,7
3	97	3	17,7	17,9	19,3	20,1
4	95	5	17,7	18,0	14,8	19,3
5	93	7	15,1	16,3	15,0	19,3
6	90	10	10,8	13,7	15,8	17,8
7	85	15	16,7	17,5	19,2	20,3

Table 3. Change of relative lengthening at breaking of the pressed samples of compositions of polythene and starch at biodecomposition

№	Structure of composition, %		$\varepsilon_{sp.}$,%	$\varepsilon_{sp.}$,%	$\varepsilon_{sp.}$,%	$\varepsilon_{sp.}$,%
	polyethylene	starch	исх.	in 14 day.	in 28 day.	in 42 day.
1	100	0	500	500	500	500
2	98,5	1,5	35	30	25	22
3	97	3	53	44	25	21
4	95	5	27	23	11	10
5	93	7	15	12	10	10
6	90	10	9	12	10	7
7	85	15	12	18	23	23

Table 4. Change MFR of the pressed samples of compositions on the basis of polythene and starch at biodecomposition

№	Structure of composition, %		MFR, г/10 mines,	MFR, г/10 mines, in 14 day	MFR, г/10 mines, in 28 day.	MFR, г/10 mines, in 42 day.
	polyethylene	starch				
1	100	0	7,26	7,1	7,24	7,18
2	98,5	1,5	50,57	60,3	77,6	89,5
3	97	3	33,3	40,6	77,2	95,7
4	95	5	65,5	78,2	134,5	193
5	93	7	100,6	135	150	193
6	90	10	120	114	158	254
7	85	15	139	97,2	83,4	88,2

The analysis of the received results showed, that at biodecomposition in ground the explosive pressure varies insignificantly whereas the relative lengthening at breaking of samples decreases. It indicates that compositions at burying in ground become more rigid as there are structural changes in the polymer's matrix, as a result of which compositions are exposed to a greater destruction, than initial polythene.

Thus, introduction of starch as an additive to a synthetic polymer allows to quicken up the process of decomposition of the polymer under the influence of microorganisms and at the same time does not have significant influence on the initial physical and chemical properties.

References

[1] Militskova, E.A., Potapov, I.I.processing of waste products of plastic - M.: *Chemistry,* 1997. - WITH. 159.

[2] Fomin, V.A., Гузеев, V.V.biodecompos polymers, a condition and prospects of use // *the Layer. Weights* - 2001, №2. - with 42-46.

[3] Sherieva, M.L., Shustov G.B. biodestroy of a composition // *Chemistry in technology and medicine:* Materials of the All-Russia scientific - practical conference. - Makhachkala, 2001. - With. 165-167.

[4] The encyclopedia of polymers / Under ред. V.A,.Kargin, etc. - M.: *the Science*, 1972. - WITH 37. T.1.

[5] The encyclopedia of polymers/under ред. V.A,.Kabanova, etc. - M.: *the Science*, 1977. - WITH 37.

In: Polymers, Polymer Blends, Polymer Composites... ISBN 1-60021-168-2
Eds: A.K. Mikitaev, M.K. Ligidov et al., pp.183-185

Chapter 25

RECEPTION AND RESEARCH OF THE PROPERTIES OF MODIFIED STARCH

Madina L. Sherieva**, Gennadi B. Shustov, Ruslan S. Mirzoev, Betal Z. Beshtoev and Inna K. Kanametova
Kabardino-Balkarian State University, Nalchik

ABSTRACT

The technologies of producing starch at the open joint-stock company «KST» (KBR, Maiskiy region, village Aleksandrovskaya). The characteristics of 3 kinds modified starch are given. The use of modified starch for receiving biodecomposable packing materials is studied.

Keywords: starch, biodecomposable polymer.

Starch is the most widely used material of all the natural compounds of biodecomposed packing materials.

Starch, as is it known, is the most widespread material of plants. Starch is formed in leaves of plants as a result of photosynthesis and is postponed in roots, tubers and seeds as grains. In industrial conditions starch is received from potato and corn. Starch of wheat, rice, sorghum and other plants has less industrial value. The production technology of starch depends on the kind of raw material and the purposes for which the starch is made.

The open joint-stock company "KSF" (KBR, Maiskiy region, village Aleksandrovskaya) 3 kinds of modified starch are produced now:

1) Starch modified for drilling.
2) Starchite.
3) Swelling food starch.

* Kabardino-Balkarian state university, Nalchik, 360004 Nalchik st. Chernyshevskaya, 173. sheri12@rambler.ru

The reception of modified starches is carried out on the Dutch rolling dryers which are warmed with steam at a certain pressure. Starch suspension of a certain density is moved on a drum rolling dryers and, having turned to paste, is dried up in a thin layer. The received film is cleaned off by a knife and goes into the crusher where through the certain apertures in the grid it is blown into the bunker to be packed in bags.

Swelling food starches that passed water and thermal treatment, get new structure, i.e. there is a splitting of polysacharide starch grains.

Received split starches have ability to swell in cold water and pass completely or partially into a soluble condition.

The technology of releasing these three kinds of modified starch is practically identical and depends only on the density of starch suspensions, chemical additives and the grid of prosowing.

1) "Starch modified for drilling" is a technical starch. For its reception into the starch suspension of 40 % C.B. salt – oxidizer and alumokaly alum ($KAl(SO_4)_2 \cdot 12H_2O$ are added, then mixed in the reactor and submitted on rolling dryer. The received film goes to a crusher with diameter of a grid 4 mm. This starch is applied as the stabilizer of clay solutions at drilling chinks in gas and a petroleum-refining industry.
2) "Starchite" is a technical starch. It is developed on the same technology, without additives, but with the increased density starch suspensions up to 42-44 % C.B. with the diameter of the cell of the grid being 5 mm. "Starchite" is applied in the foundry industry as a forming material while manufacturing pastes, i.e. it is used as a softener and holder of superfluous moisture of forming mixes at work on automatic transfer lines for casting blocks of automobile engines.
3) "Swelling food starch" is a food starch. It is developed also without additives, but with the density of starch suspensions lowered up to 36-38 % C.B. and prosowing through a sieve with diameter of a cell of 3 mm. This starch is applied in the various food-producing industries as an additive to condense mayonnaise, ketchup, tomato paste, jam, ice-cream, etc., it is also used to improve the quality of flour instead of gluten (5 kilos per 1 ton of flour). This starch is used for producing puddings of fast preparation, for producing protein-free food stuffs as bread, macaroni, etc. It is also widely used for briquetting forage; agglomeration of various products such as powder, ores, coal, etc.

The quality of these three kinds of modified starch is according to their ability of swelling, holding of superfluous moisture and stabilizing ability of viscosity and solubility and is regulated by specifications on each kind of production.

The simultaneous application of softeners and glycerin allows to receive flexible thermoplastics of starch using compressive pressing [2] and extrusion [3].

The materials received from corn, potato and wheat starch, containing constant in relation to starch quantityof glycerin (1:0,3) and from 8 up to 25 мас. % of water, were elastic, i.e. had T_c below 20 C . While researching mechanical properties of such materials the dependence of the module of elasticity of an explosive pressure of samples not only on the contents of the softeners, but also on the nature of the starch was found out. In the opinion of the authors of the work [3], the reason of difference of mechanical properties can be caused by the big maintenance of amylopectin in the corn starch which is better masticated with water, than potato starch enriched with high-molecular amylose.

On the of the invention [4] we received samples destroyed starch containing 80 % starchite, 1 %, hydrogenated fat and 18 % of water. After preparation the mixture has the form of a loose powder. The received mixture is loaded into the plodder, in the auger cylinder (temperature ≈160-170 C) the given powder fuses. Then the fusion is pressed through and divided into grains with average diameter 2,5-3,0 mm. The material has a form of a firm white product with thin foamlike structure. With the received material it is possible to press test samples, suitable for studying their properties. The durability and flexibility of the received samples can be noted. It should be noted, that if to press test samples at once from the prepared composition they turn out to be more fragile.

Besides glycerin and polyglycols, plasticizing effect on starch has such substances as sorbite, natrium salt of dairy acid, urea, ethylene-, diethylene-, polyethyleneglycol and diacetate glycerin. [5]

Water used in extrusion starch does not only transfer the system into the thermoplastic condition, but also partially protects the polymer from destruction. Addition of water and others hydroxide-containing substances are used for disposable or not long-term application. In this connection mixes of starch with synthetic polymers get the increasing value. These materials combine properties of the synthetic component present in them and have the ability of biodegradation due to the presence of a natural biodecomposed component - starch in the system [6-9]. It is necessary to note, that biodecomposition of films with similar structures (on method astm-d-5209-92) occurs actively with allocation of $_{co2}$, microbiological weights and the metabolic products useful to plants [7].

REFERENCES

[1] Gajria, A.M. et. al. // *Polymer*. - 1996. - V.37, №3. - p.437-444.

[2] Hulleman, S.H.D., Janssen, F.H.P., Feil, H. // *Polimer*. - 1998. - V. 39. - P. 2043.

[3] Della Valle, G., Bullen, A., Carrean, P.J., Lavoie, P.-A.-, Vergnes, B.// *J. Heal*. - 1998. - v. 42. - p. 507.

[4] *The Description of the Invention to the Patent of the USSR 1612999*, with 08 1 3/02. a way of formation products from compositions on the basis of starch.

[5] Lourolin, D., Coighard, L., Bozot, H., p.colonna // *Polimer*. - 1997. - v. 38.-p. 5401.

[6] Sherieva, M.L., Shustov, G.B., Shetov, R.A.. biodecomposed{biodecayed} compositions on the basis of starch // *the layer. weights* - 2004. - №10. - c. 35-39. the patent the usa 5498692, (1996).

[7] Liu, W., Wang, Y.-J., Sun, Z. effects of polyethylene-graphted maleic anhydride (pe-g-ma) on thermal properties, morphology, and tensile properties of low-density polyethylene (ldpe) and corn starch blends. // *J. Appl. Polym. Sci*. - 2003. - v.88, № 13. - p. 2904-2911.

[8] Long, Y., Yeo, Ch. G. B., et al. *Biodegradation Polymer. A Stalemate*. 753328 (australia, мпк 6 with 08 1 003/06, c 08 k 005/09). заявл. 13.12.1999. опубл. 17.10.2002.

In: Polymers, Polymer Blends, Polymer Composites… ISBN 1-60021-168-2
Eds: A.K. Mikitaev, M.K. Ligidov et al., pp.187-191

Chapter 26

BIOLOGICALLY UTILIZED PLASTICS: CONDITION AND PROSPECTS

Gennadi B. Shustov, Madina L. Sherieva*, Ruslan S. Mirzoev, Inna K. Kanametova and Betal Z. Beshtoev
Kabardino-Balkarian State University, Nalchik

ABSTRACT

The survey of literary sources devoted to the problem of creation of biologically utilizable plastics. In the sake of creating biodecomposable materials the researchers turned to various of row material of both synthetic and natural origin. Ecological consequences of introduction of biodecomposable polymers are considered.

Keywords: survey of literary, biodecomposable materials, Ecological consequences.

The packing material, polymeric film, polythene (polyolefins, ethylene copolymers, etc.) have received such a wide application, that neither human activity, nor, all the more, natural environment is capable to cope with the inflow of polymeric waste products. In this connection there appeared a necessity of manufacturing of polymeric materials capable of biodecomposition under the influence of the environment and microorganisms.

Now there is a lot of such developments and ideas of creation of such polymeric materials, but they either are insufficiently developed, or are not effective in the economic way.

For creation of high-quality and economic bioplastics researchers turned to various sources of raw material, such, as corn starch, capola, castor oil, soya fiber and so on. Plastics polyols on the basis of soy bean used for carpet coverings are already developed and produced now. Technologies for producing bioplastics on the basis of soy oil, and also new

*Kabardino-Balkarian state university, Nalchik. 360004 Nalchik st. Chernyshevskaya, 173. sheri12@rambler.ru

biotechnologies with application of fermentative processes with the purpose of use of animal fats, vegetable oils and industrial wastes are developed [1].

The current situation on development and mastering of biodecomposed polymers is given estimation and three basic directions in this area are allocated: ПЭФ hydroxide carbonic acids (glycollic, dairy, valerianic), plastics on the basis of reproduced natural polymers (water-soluble ПЛ from a mix of starch and pectin, and also a mix of starch from PVC, ПВС, compositions are actively developed on the basis of cellulose, chitin), giving biodecompositionability to industrial high-molecular synthetic materials (ПЭ, software, PVC, ПС, ПЭТФ). Now three directions of giving biodecompositionability to large-tonnage polymers are developed: getting of compositions with biodecomposed natural additives (compositions ПЭ with starch, etc. Biodecomposed additives), the directed synthesis of biodecomposed plastics on the basis of industrially - mastered synthetic products (synthesis corresponding ПЭФ and polyetheramides), introducing into the structure of biodecomposed polymers of the molecules containing in their structure functional groups, promoting the accelerated photodecomposition of polymer (СПЛ ethylene with carbon oxide, introduction of vinylcetone monomer as СПЛ ethylene or styrene, introduction in ON dithiocarbamic iron, nickel or corresponding heroxides, and also introduction of a pulp of cellulose, alkylcetones or fragments containing carbonyl groups) [2].

Synthetic Polymers

Most frequently starch is used to modify polythene (ПЭ) - a film material which is usually used for short-term application. Thermo-softening mixes of synthetic polymer with starch are received by using, as a rule, starch, plasticized glycerin and water. Biodecomposition is promoted usually by use of additives of small quantities of prooxidizers. For example, such composition is: ПЭ - starch - vegetable oil. [3]

At the 16 congress by D.I.Mendeleyev .I.Suvorova and research assistants presented a biodecomposed mix of starch and synthetic polymer. The biodecomposed materials received on the basis of renewed raw material were presented. The properties of mixes of starch with hydroxypropyl-, carboxymethyl-and methyl - Ц, polyethyleneoxides and copolyamides. The sorbtion and diffusion of water steam and absorption of water in such materials were investigated defined. Properties of the materials were investigated by ГХ methods with detecting on heat conductivity at using He as the carrier. The speed of biodecomposition was defined according to the speed of allocation $_{CO2}$, the dependence of biodecompositionability from the maintenance of starch in the material was investigated [4].

For the convenience of mulching the films are received from polyolefin introducing into the composition of photosensitive additives - iron and nickel dithiocarbamat or corresponding heroxides. With the purpose of acceleration biodecomposition of films on the basis of polythene for an agriculture, polypropylene pulp of cellulose is put into them [5].

Cellulose, starch, polyethers in this case are a source of a nutrient medium for microorganisms due to that there is an almost compete biodecomposition.

But these development are insufficiently effective, as synthetic polymers are exposed to biodecomposition very badly.

In this connection new technologies of creation of biodecomposed polymers were developed, these are polymers are of a natural origin, such as pectin, cellulose, starch and others. With addition in KM chitosan ПЛ with improved superficial properties at preservation of ability to biodecomposition is received. It is shown, that when composting destruction of ПВС begins. [5, 6].

POLYMERS OF A NATURAL ORIGIN

Plastics of soy protein and corn starch, made by various methods, are investigated on bio destroyability on kinetics of CO2 allocation. Molded materials made of them are exposed to biodecomposition faster than raw materials. This effect is charged due to denaturation of protein and gelatinization of starch. At reception of plastics soy protein, corn starch, softeners were mixed and samples were mould.

In 2003 in Vladimir at the scientific and technical conference the material of creation biodecomposed film nanoaggregate on the basis of cellulose and starch were presented. Liquid nanoaggregate solutions of cellulose in MMO represent discrete box-cover structures of clay into the interbatch spaces which macromolecules of cellulose forming with a polymeric matrix labile structural associates are included. The revealed structural transformations predetermine also the operational properties of nanoaggregate films on the basis of cellulose, starch and natural, layered silicates. Introduction of starch into the cellulose solutions allows to receive a new film material with more than 4 times increased, in comparison with cellulose films, moisture-holding properties [5].

The compositions (KM) containing starch (KP), polyvinyl spirit (ПВС) and glycerin cast from a solution in the SQUARE. At composting ПЛ within 45 day KP and glycerin completely decay, whereas ПВС remains basically not destroyed. ПЛ from KM with maintenance of ПВС of 20 % are determined as having required physical characteristics at maintenance of KP in quantities sufficient for biodecomposition. While adding chitosan into KM ПЛ with the improved superficial properties at preservation of ability to biodecomposition is received [6].

At the IX All-Russia student's scientific conference devoted to the 130-anniversary of opening of D.I.Mendeleyev's Periodic law in Ekaterinburg, the report on phase division in a biodecomposed mix of starch and polymer with vinyl acetate was presented. [7]

With the method of hot formation under pressure films of mixes of starch with a сэвиленом-copolymer vinyl acetate (BA) with various maintenance of BA in a copolymer (from 5 up to 25 %). are received. With the help of water sorbtion in a liquid phase it is revealed, that with the increase of maintenance ofBA in the mixture water absorption increases. The method of points of turbidity phase diagrams of mixes are received and is shown, that with the increase of maintenance of BA in the system miscibility of сэвилена with starch improves. Biodegradability was estimated with the method of gas chromatography by comparison of speed of allocation of carbonic gas at biodegradation films in watersoil suspension. It is revealed, that speed of biodegradation grows at the increase of maintenance of starch in a polymeric composition. The received data allow to choose optimum structure of components of the investigated mixes, providing good operational properties and ability to biodecomposition [7, 11].

The world's largest factory producing packings for foodstuff by a method of polymerization of a dairy acid of 140 thousand ton per year , plastics biologically was started in operation in the USA. Films from a mix of corn starch and a polymeric dairy acid, extruded corn flour etc. can be used as a material for manufacturing such packings as well. It is marked, that these materials are similar to synthetic polymers. The firms which produce the given materials are listed, and it is pointed, that the basic lack of these materials is their high cost which more than 2 times exceeds the cost of polystyrene and polypropylene, and in the USA and Europe factories producing new packing materials biologically decomposed after using according to their purpose have already been started.[8].

Firms Petroplast AG and Vinora AG (both in Switzerland) are engaged intensively in the search of packing materials which are alternatives to polythene and are destroyed biologically. To this materials belong those grown from raw material, first of all on the basis of corn starch - CompoBag with use of product Mater-Bi of firm Novamont belonging to chemical group Montedison (Italy). This product is processed, as well as traditional polymers, is painted biologically destroyed by uterus mixes or natural pigments, thermally vignetted on a paper, cardboard, a cotton and other natural fabrics, is antistatic, is sterilized and sticked together with traditional glues. Some types Mater-Bi can be used while producing packing materials for food stuffs [9].

ECOLOGICAL CONSEQUENCES OF INTRODUCTION OF BIODECOMPOSED POLYMERS

There is an assumption in the literature, that biodecomposed polymers brought into ground can negatively influence the growth of plants. Therefore biodecomposition of starch (КХР), straw, polyhydroxybutyrate, polylactide and thermally processed and mixed ПЭ up to a biomass and CO_2, and also their influence on the growth of watercress (КС) and millet is investigated in the work. The change of pH, volatility, breath of the ground, the maintenance of metals were studied. Insignificant influence КХР on the growth of plants, strong microbiological decomposition of straw without chemical-toxic separations with some initial delay of growth of КС is established. Some biodecomposed polymers cause insignificant delay of growth, but then it is normalized. It is established, that there is no observed connections between change of concentration of ions, pH-quantities with the factor of the slowed down growth of plants; the process has no chemical basis [10].

To sum up, it is necessary to note, that yet we can not do without polymeric materials which are in the lead on the degree of environmental contamination, but using traditional plastics means ignoring the fact, that after any processing they sooner or later appear garbage with which neither people, nor nature can do anything. Therefore only the use of decomposable polymeric materials is the reasonable alternative in preservation both the planet and health of its inhabitants.

REFERENCES

[1] Crandall, L. Bioplastics: A burgeoning industry // *INFORM: Int. News Fats, Oils and Relat. Mater.* - 2002.13, N 8. - P. 626-627, 629-630.

[2] Fomin, V.A., Гузеев Century. Century. Biodecomposed{biodecayed} polymers, a condition and prospects of use // *Пластич. Weights* - 2001, № 2. - WITH. 42-48.

[3] Suvorova, A.I., Tjukova, I.S., Труфанова E.I.biodecompos polymeric materials on the basis of starch // *Successes of chemistry.* - 2000. - № 5. - With. 69.

[4] Suvorova, A. I., Tijkova, I. S., Truvanova, E. I. Biodecomposed{biodecayed} mixes starch/synthetic polymers. Biodegradable starch / synthetic polymer blends // 16 Mendeleev Congress on General and Applied Chemistry, Moscow, 1998. Vol. 2. The Present State-of-Art and the Development of Chemical Production. Materials for Future and Non-Traditional Chemical Technologies. *Chemical Sources of Energy.* - M.: Publishing house IOH of the Russian Academy of Science, 1998. - With 458.

[5] Head, L.K., Kuznetsova, L.K., Queens, Ю. M., Куличихин Century.. Biodecomposed{biodecayed} film нанокомпозиты on the basis of cellulose and starch. Ethers of cellulose and starch: synthesis, properties, application // Materials 10 anniversary All-Russia scientific and technical conferences with the international participation, devoted to the 45-anniversary of creation of a scientific direction " *Ethers of cellulose* ". - Vladimir: Посад, 2003. - With. 287-290.

[6] Jayasekara, Ranjith, Harding, Ian, Bowater, Ian, Christie, Gregor B. Y., Lonergan, Greg T. Biodegradation by composting of surface modified starch and PVA blended films // *J. Polym. and Environ.* - 2003. V. 11, N 2. - P. 49-56.

[7] Toropova, S.M., Butorina, E.J., Trufanova, E.I., Tjukova, I.S., Суворова, A.I. Fazovoe division in biodecomposed{biodecayed} mixes of starch with copolymers этилена with винилацетатом. Problems of theoretical and experimental chemistry // *Тез. докл. IX Всерос. студ. науч. конф., посв. To the 130-anniversary of opening of the Periodic law of D.I.Mendeleyev.* - Ekaterinburg: Publishing house UrGu, 1999. - With. 224.

[8] Caranti№S. Materiaux plastiques biodegradables. \\ *un decollage timide! Rev. lait. fr.* - 2003. - N 635. - WITH. 31-32, 34.

[9] Folien aus biologisch abbaubaren. // *Werkstoffen Coating.* - 2002. V.35, N 4. - P. 120-121.

[10] Okologische Auswirkungen des Einsatzes biologisch abbaubarer Materialien in der Landwirtschaft // *Osterr. Chem.-Z.* - 2003. - B.104, N 4. - S. 13-14.

[11] Syguchova, O.V., Kolesnikov, N.N., Lihachov, A.N., A.A.role's Priests крахмального a component in process деструкции mixes with ЭВА-ТПК at influence плесневых mushrooms // *Пластич. Weights* - 2004. - №9. - with 29-32.

In: Polymers, Polymer Blends, Polymer Composites... ISBN 1-60021-168-2
Eds: A.K. Mikitaev, M.K. Ligidov et al., pp.193-195

Chapter 27

COMPOSITE MATERIALS CAPABLE OF MULTIPLE PROCESSING (ECOLOGICAL ASPECTS OF THE PROBLEM)

A. Yu. Bedanokov, O. B. Lednev, A. H. Shaov, A. M. Kharaev and B. Z. Beshtoev
The Maykop State Technological Institute
The Kabardian-Balkar State University

At the end of the 20th century, which is often called as the century of polymers, we can be firmly convinced that the future of the national economy will be defined by creating and using new materials. Possessing the set of valuable characteristics such as high durability, little weight, flexibility, specific electrical properties, chemical stability, to the fast and mass production and processing into the items of complicated form and different colours the polymers took the first place practically in all branches of production.

However expansions of production and use of polymeric materials raised before mankind a problem of placing their wastes and repeated use of worked plastic materials. In Russia this problem hasn't been discussed though the struggle for keeping the Earth from littering plastic wastes is going on all over the world. The rational ways of using of polymeric wastes is constantly being developed.

It is known that the information is growing at least twice as fast that industrial potential. Nowadays more that 10.000 of periodical editions reflecting the ecological themes are appeared in the world. A number of organizations examining these problems several times as much. That is why one of the important tasks is the realization of information support of ecological researches (particularly the researching in the field of polymeric wastes utilization).

At present there are 4 trends of process of the plastic wastes utilization:

1) Recycle in materials.
2) Chemical way to the getting original raw materials and pyrolysis.
3) Burying of biodegradable polymers.

4) Burning.

The creation of such polymeric materials which are capable of multiple processing, reserving at high level exploiting characteristics is considered to be a base of successful realization of material recycle of polymeric wastes.

Polydefins particularly polyethylene of high density (HDPE) belong to the class of thermoplastics which can be used in the various fields of engineering. Every day tons of polymers all over the country are thrown away as waste products (mainly wrappings and packaging). It is known that polyethylene can pollute the environment without biodestruction for a long time. That is why the examining of high-density polyethylene utilization problem is guite actual. With that purpose we were examining the molecular weight change (M_w) of with the multiple extrudering (n=1-5), and also the character of phosphoroorganic polymer influence on the molecular weight of polyethylene.

As it is known, polymers are not used without addional stabilization. As polyolefin's stabilizers particularly for HDPE various phenols with tretbutul substitutes are often used. One of them is Irganox-1010 (Swiss production). In this connection one of purposes of our testing was the comparison of influence character of phosphoroorganic polymer and Irganox-1010 on M_w change with multiple processing of HDPE.

Phosphoroorganic polymers synthesized by low temperatured acceptor-catalyst polycondensation of diphenilolpropane with methyldichlorphosphanate and has viscosity 0,4дл/г (dichlorethane; T=293K; c = 0,5 г/дл.

The compositions of polyethylene with phosphoroorganic polymer (0,05 - 0,5%) were prepared by the method of exstruding the blend of initial components (T= 473K; 10 - 12 об/мин; the length of heating part of extruder is 22см)

Melt index, characterizing rheological properties of polymer melts for HDPE and compositions on its base were determined (IIRT-M type) at 463K and2,16 and21,6 kg Load (Russian normative quality document 11645-73), and calculating was done by the following formule:

$$ПТР=(m_{cp}\times\tau_0)/\tau ,$$

where ПТР-melt index;

τ_0=600c-standart testing time for polyethylene;

τ -time of melt outflow in the experiment;

m_{cp}= average weight of three measurements.

The values of M_w (molecular weight), M_n (molecular-mass distribution) were calculating on the base of melt index data using known ratio for HDPE:

$$\lg M_w = \lg 129000 - 0{,}263\times\lg ПТР^{463}_{2,16}$$

$$\lg (M_w/M_n)=\lg 0{,}0275+1{,}4\times\lg \left(ПТР^{463}_{2,16} \Big/ ПТР^{463}_{21,6} \right),$$

where $ПТР^{463}_{2,16}$ - melt index value at 463K and 2,16kg load

$\Pi TP^{463}_{2,16}$ - melt index value at 463K and 21,6kg load.

Usually (M_n) value characterizes alow-molecular part of (MMD), and (M_w) value characterizes a high- molecular part of (MMD) it is determined that M_w HDPE increases from 269000 till 303000 after single extruding and falls down to 214000 ofter quintuple extruding. Addaning phosphoroorganic polymer in quantity 0,5 allows to kup molecular mass of HDPE practically at the same level (295000) despite extruding division (up to 5 times). Industrial polyolefin stabilizer Irganox-1010 maintais M_w HDPE within the limits 245000-257000. It is necessary to emphasize that the presence of phosphoroorganic polymer in polyethylene makes the polymer fireproof better.

These results of testing tell us about the perspective using of phosphoroorganic polymer stabilizer and modifier during the utilization of wastes of nigh density polyethylene.

REFERENCES

[1] Govariker, V., Visvanathan, N., Schridhar, J. *Polymers*. Moscow. Science, 1990 396 pages.

[2] Achoh, S., Achoh, S. The role of information in the organization of environment. The materials of the 3[d] scientifis -proctical conference of the Maykop state Technological Institute. *Ecology and Forestry*.- Maykop 1998-132 pages.

[3] Militzkova, E., Popov, I. *Processing of wastes of plastics*. - Moscow -1997, 159 pages.

[4] Mascukov, N., *Stabilization and modification of high polyethylene by oxygen acceptors*: Theses of doctors of chemistry- Moscow, 1991-422 pages.

In: Polymers, Polymer Blends, Polymer Composites... ISBN 1-60021-168-2
Eds: A.K. Mikitaev, M.K. Ligidov et al., pp.197-200

Chapter 28

ECOLOGICAL AND ECONOMICAL ASPECTS OF COMPOSITION MATERIALS CREATION

A. Yu. Bedanokov, I. V. Dolbin, A. H. Shaov, A. M. Kharaev, B. Z. Beshtoev and A. K. Mikitaev

The Center of State Sanitary Inspection, Maykop;
Kabardino-Balkar State University, Nalchik

Pesticides are considered to be the most dangerous of all chemical compounds which are received with air, water and food by human organism. In 1988 the USA. National Academy of Sciences published a report which noticed that more than one million Americans risk to fall ill with cancer caused by twenty-eight cancerogen pesticides in food. The abuse of pesticides can provoke a burst of cancer diseases and mutations in developing countries. According to the World Health Services Organisation data 500,000 people are poisoned by pesticides and die yearly.

Chlororganic pesticides are widely used in agriculture (against vermin), forestry, veterinary and medicine. These compounds are characterized by two very important properties. Firstly, it is firmness to environment factor influence such as temperature, solar radiation, moisture. Secondly, they are expressed cumulative properties. All this caused a situation that firm chlororganic pesticides are found almost in all living organisms and their concentration in tissue and organs is high than in the environment.

DDT (dichlor diphenil trichloroethane, dichlordiphenil trichlormethyl-methane) is a chlororganic pesticide. It is a white crystal substance without taste and smell. In 1956 its world production was 80,000 tons. From 1942 till 1974 4,5 million tons of DDT were spent for agricultural vermin destruction. However, the World Health Services Organization forbade using that preparation because of its toxic influence on human organism through "food chain", consisting of plants, animals used by human being.

Despite the taken measures, the problem of influence of a given compound on environment and human organism is still vital. Firstly, DDT is firm to decomposition (is stands heating up to 115-120°C for 15 hours and doesn't decay at cooking) and can circulate in biosphere more than fifty years. Also, it is absorbed easily in deposits and soils which can be a depot for DDT and its derivatives. Such depots are the sources of chronic influence. Secondly, this given preparation is still used in developing countries. Thirdly, the DDT spreading has a global character. It can be transported in migrating animals organisms, and with air and ocean streams too. That is why the DDT influence on environment is much wider than its using region (for instance, on Antarctic Continent, very far from using zones more than 2000 tons of DDT were accumulated in glaciers).

The problems of ecological safety became aggravated. It is caused by unstable social and economical conditions in Russia. In particular, the problems connected with processing and destroying of unused pesticides are solved very slowly and uneffectively. By the first of January 1998 11,75 tons of pesticides and more than 2,000 tons of DDT had been accumulated in the warehouses of the Republic of Adygea. The lack of utilization ranges forced the farms of the Republic to spents means on safe keeping of these preparations. The need in warehouses for long duration keeping of pesticides was satisfied only at 52% but the quantity of unused chemical preparations increases every year.

Thus it turned out that the situation is neither ecologically not economically profitable. It is a result of the unsufficiently considered agricultural activity.

One of the ways of solving this problem is the elaboration of the processing methods for unused and worthless pesticides.

We have worked out the DDT utilization method. It is the creating of composition materials. The method of synthesis of dichlorbenzophenon (dichlordiphenilketone) from DDT (that is alkaline dehydrochlorination of dichlordiphenil trichlorethane in ethanol with following oxidation by chromic anhydride in icy vinegar acid. On the basis of the obtained dichlorbenzophenon and diphenylolpropane as well as dichlorbenzophenon and phenolphtalane by the method of high-temperature polycondensation in surroundings of dimethylsulphoxide in nitrogen atmosphere the oligoketones of dian and phenoiphtalane rows with degrees of condensation 1.5.10.20. The synthesis of oligoketones was made at mole surplus of diphenylolpropane or phenolphtalane to dichlordiphenilketone according to the following scheme:

The structure of obtained oligoketones was confirmed by IR-spectroscopy and definition of hydroxide group quantity. The availability of absorption stripes in IR-spectrums corresponding to simple ether joints in domain 1135cant^{-1}, to isopropyliden group at dian surplus 2960 - 2980cant^{-1} (in the case of dian oligorners), to lactone group 1710 - 1760cant^{-1} (in the case of phenolphtalane oligomers), hydroxide group 3300 - 3600cant^{-1} and keto-group 1600 167Scant^{-1} testifies of oligoketones formation.

Some properties of oligoketones are cited in Table 1.

Synthesized oligoketones were researched as HOPE modifiers. To evaluating the effectiveness of putting the oligomers into the HDPE melt, their 0,1% (at mass) concentration was studied.

The experimental plant (with the help of which all physics and mechanic properties of compositions on the basis of HDPE and aromatic oligomers) represents a pendulum setting UT-1/4 which is supplied with sensor of loading. Its signal was transmitted directly to memory oscillograph, model C 8-12.

It was found out that oligoketones on the basis of diphenylolpropane, independently from condensation degree, influence on HDPE as plasticizers. We can make a conclusion on these facts that there is some increasing of relative deformation during the destroying of models. But at the transition to phenolphthalein oligoketones there is an essential distinction, that is the polymer models become harder (the modulus of elasticity increases). Both the shock viscosity and the limit of forced stretchiness stay at high level which exceeds a little these parameters for primary polyethilene. Apparently, it is connected with harder structure of the phenolphtalane in compare with dian. Practically all researched oligoketones increase cristallinity degree of the polyethilene. Probably, getting into amorphous part of the polymer can become the "embryo" of crystallization. During multiple (five times) extruding the character of oligoketones on physical and mechanic properties of high-density polyethilene doesn't change.

Table 1. The properties of aromatic oligoketones

Oligoketones	Output, %	$T_{softing}$, K	M.M.	OH-group content, %	
				Calculated	Found
OK- 1D*	98	402-408	634,78	5,36	5,30
OK-5D	98	420-425	2260,72	1,50	1,55
OK-10D	99	433-438 4293,17	4293,17	0,79	0,75
OK-20D	99	440-448	8358,43	0,41	0,40
OK-1F**	98	469-473	814,85	4,17	4,20
OK-5F	98	483-488	2800,94	1,21	1,20
OK-10F	99	510-517	5283,75	0,64	0,65
OK-20F	99	528-533	10248,77	0,33	0,60

D*-oligoketones on the basis of diphenylolpropane with condensation degrees 1-20. F** - oligoketones on the basis of phenolphthalein with condensation degrees 1-20

Besides physic and mechanic properties of obtained compositions we have researched such characteristics as therrnostability, melt index and molecular-mass distribution, chemical firmness, dielectric properties.

Complex study of aromatic oligoketones' influence on propertiesof hign density polyethilene allows to recommend them as quite perspective modificators ofHDPE. "

Also, we have to notice, that the represented oligoketones in the capacity of additive to high-density polyethilene can be used as fbrpolyrners for the synthesis of high-molecular compounds of aromatic polyetherketones class which are very perspective materials of construction purpose with higher physics and chemical characteristics.

Thus, the results of our researches show, that with direct utilization of DDT we san obtain new perspective composition materials. This method permits to solve a very vital ecological and economical problem of processing of unused chlororganic pesticides.

REFERENCES

[1] Novikov, U.V. Ecology, Environment and Human Being. Moscow.:*FAIR*, 1998.-320p.

[2] Korshak, B.B., Rusanov, A.L. Thermo-fire firm polymers on the basis of chlor and its derivatives. *Chemistry achievments,* 1989, T.LVIII.C1006.

[3] Melnikov, N.N., Nabokov, V.A., Pokrovsky, E.A. DDT - *Properties and Using.* M.. 1954.

[4] Hygiene criteria of environment condition DDT and its derivatives: ecological aspects. *World Health Services Organisation*. Geneva, 1991.

[5] Poller, Z. Chemistry on the way to the third millennium. Moscow.: World, 1982. Sheugen, A.H., Tkhakushinov, A.K., Kozmenko, G.G. *Recreation resources of Adygea*. Maikop.: Adygea, 1999. - 272p.

[6] Shaov, A.Kh., Kharaev, A.M., Mikitaev, A.K. General methods of obtaining of the $4,4^1$-digalogenbenzophenones - monomers for the synthesis of polyetherketones (review). *Plastics,* 1990, ©12. -P.35-38.

In: Polymers, Polymer Blends, Polymer Composites... ISBN 1-60021-168-2
Eds: A.K. Mikitaev, M.K. Ligidov et al., pp.201-206

Chapter 29

POLYARYLATE OXIMATES (PAO), THEIR PHYSICOCHEMICAL PROPERTIES AND STABILIZING INFLUENCE ON POLYALKYLENE TEREPHTHALATE (PAT)

Yu. I. Musaev[2], A. M. Kharaev, E. B. Musaeva[2], V. A. Kvashin[2], A. B. Dzaekmukhove, M. A. Mikitaev[1], A. I. Eid [1] and Yu. V.Korshak[1]

[1]D.I. Mendeleev University of Chemical Technology of Russia, Miusskaya Pl., 9, Moscow 125047, Russia

[2]H.M.Berbekov Kabardino-Balkarian State University, Chernyshevsky Str., 173, Nalchik 360004, Russia

ABSTRACT

Physical and chemical properties of polyarylate oximates (PAO) synthesized by catalytic-acceptor polyesterification from terephthalic acid anhydrides and diacetylphenyloxid are submitted. The modification of polybutylene terephthalates (PBT) by PAO in amount of 0,5-1 %wt. increases its thermal stability and heat resistant.

Key words: polyarylates; polyesters; antioxidants; thermal stability; polybutylene terephthalate.

INTRODUCTION

Polyarylates represent a rather perspective class of polymers, which can be successfully applied in many fields of polymeric technology required a use of materials with high heat-resistance, good dielectric and mechanical properties [1]. In industrial scale polyarylates are produced on the basis of various diphenols [2].

Experimental Part

Polymer Preparation

Polyarylate oximates (PAO) were prepared from various diketoximes (DKO) and dichloride anhydrides of *iso*-phthalic (CAIP) and terephthalic acids (CAT). Synthesis of polyarylate oximates was carried out by low temperature catalytic-acceptor polycondensation, with triethylamine as catalyst (Fig.1)

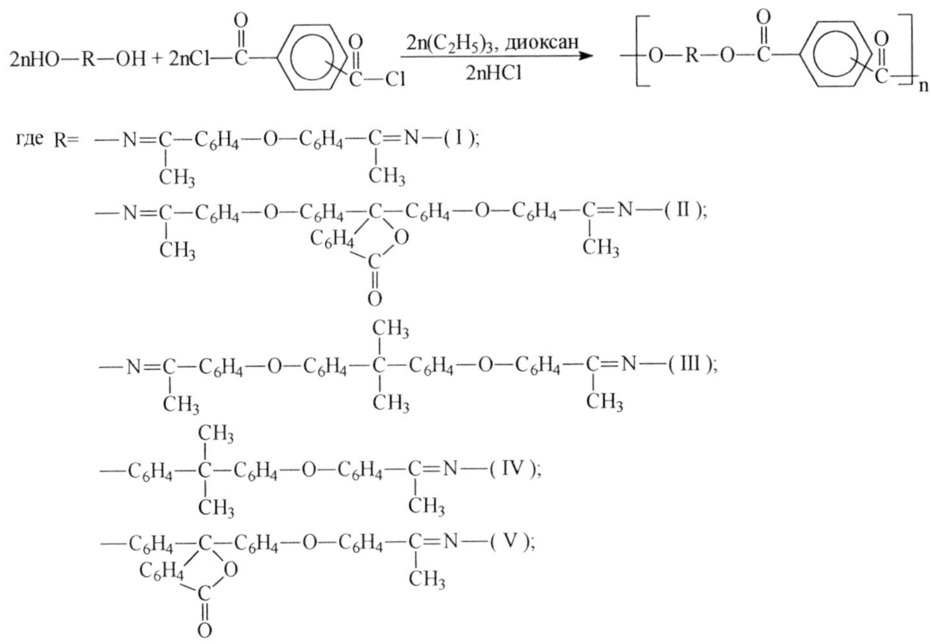

Fig. 1. Schematic representation of polyarylate oximates synthesis

The full-scale test were performed to find the influence of reaction time, solvent nature, temperature, and reagents feed ration on intrinsic viscosity of polyarylate oximates to find out the optimal reaction conditions for their preparation in accordance with the Scheme 1. It was found that synthesis of polyarylate oximates can be carried out more efficiently in dioxane at 303 K during 40 min. with triethylamine as catalyst and molar ratio DKO/ (CAT) / (DCAIP) = 2/1/1. The reduced specific viscosity η_{red} for a resulting polyarylate oximate was within the range of 0.71-0.81 dl/g. It should be noted that synthesis of polyarylate oximates from phenyl ketoximes in acetone as a solvent proceeded as a heterogeneous process and the resulting polymer precipitated from a solution.

Chemical structure of prepared polyarylate oximates was supported by elemental analyasis data and by IR-spectroscopy (Table 1). A group of bands in the region of (1735 – 1750 cm^{-1}) and the absence of absorption at 3300 – 3600 cm^{-1} belonging to hydroxyl groups is an argument in favor of polyarylate oximate structure formation.

Table 1. Reduced viscosity and elemental analysis data for prepared polyarylate oximates

Polymer	η_{red} tetrachloroethylene/phenol	Elemental analysis data *		
		C,%	H,%	N,%
PAO-I	0,75	*64,73* 65,98	*3,96* 4,18	*13,45* 14,07
PAO -II	0,81	*73,59* 74,05	*4,15* 4,28	*3,57* 4,18
PAO -III	0,71	*75,01* 75,17	*5,00* 5,29	*4,39* 4,83
PAO -IV	0,68	*76,85* 77,02	*5,03* 5,18	*2,81* 2,90
PAO -V	0,80	*73,43* 73,68	*4,14* 4,21	*2,31* 2,46

* Numbers: numerator — found; denominator – calculated.

Methods and Investigation

A number of physical methods were used for to characterize properties of the prepared polymers. The degree of crystallinity was determined by X-ray diffractometer DRON-6.0 using nickel-filtered radiation CuK_{α} (1.5405 Å). A sample was exposed within the θ angle range from 7 to 45 degrees with preset exposure spacing of 1°/min. and measurement accuracy of 0.030 degree. Chemical stability of polymers was studied on disk-like film samples with diameter 5×10^{-3}m by measuring a change of their weight during exposition according to GOST № 12020-72. Thermomechanical properties were studied by applying UIP-70 device with a constant stress of compression 0.08 MPa. The softening temperature of a polymer was found as a point at which the tangents of two branches of thermo-mechanical curve intersect. The differential scanning calorimetry (DSC) was performed in argon atmosphere by the Mettler TA – 4000 instrument supplied by DSC-30 cell with a heating rate of 20 C°/min. Dielectric strength of the obtained polymers were determined by using the high voltage generator BMW-30-01 as a part of AM-A-02F1 analyzer. Thermo-gravimetric analysis was carried for 25 mg samples in the air by using "MOM "(Hungary) instrument with a heating rate of 5 C°/ min.

Results and Discussion

The results of X-ray analysis of synthesized polyarylate oximates show the high crystallinity in samples that, perhaps, may explain bad solubility of these polymers in chlorinated solvents (chloroform, dichloroethane, dicholoromethane). The examined samples of PAO showed a good chemical resistance against the influence of aggressive medium such as H_2SO_4 (of 10 % and 30 % concentration), concentrated HCl, and NaOH (of 10 % and 50 % concentration), which was measured according to GOST 12020-72 by measuring a change in weight of samples (the amount of the extracted substances). The diluted solutions of acids had no actual effect on PAO that was in agreement with the observed absence of significant change in the weight of samples during 24 hours, and small loss in the weight (not exceeding 2 %) after 28 days. At the same time PAO displayed less stability in alkaline solutions. In the concentrated sulfuric acid samples of all synthesized polymers were dissolved after 1 day.

The character of thermomechanical curve of PAO showed that the tested samples revealed a rather rigid structure with high glass and viscous flow temperatures. The values of glass and viscous flow temperatures from thermomechanical data (Fig. 2) were found to be around 393 K and 468 K, respectively.

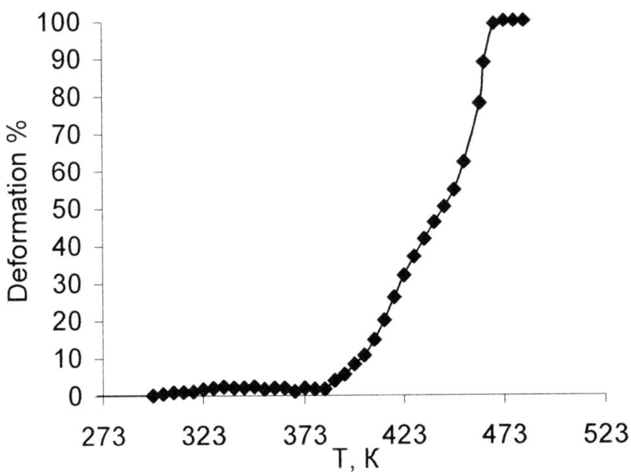

Fig. 2. Thermomechanical curve for polyarylate oximate PAO-1

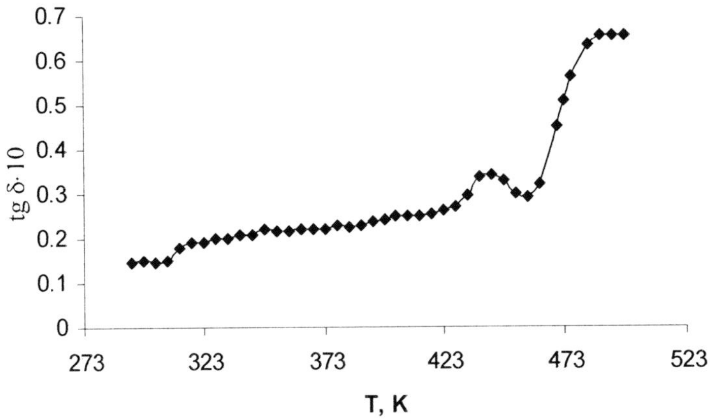

Fig. 3. A plot of loss-angle tangent (tgδ) versus temperature for PAO-1

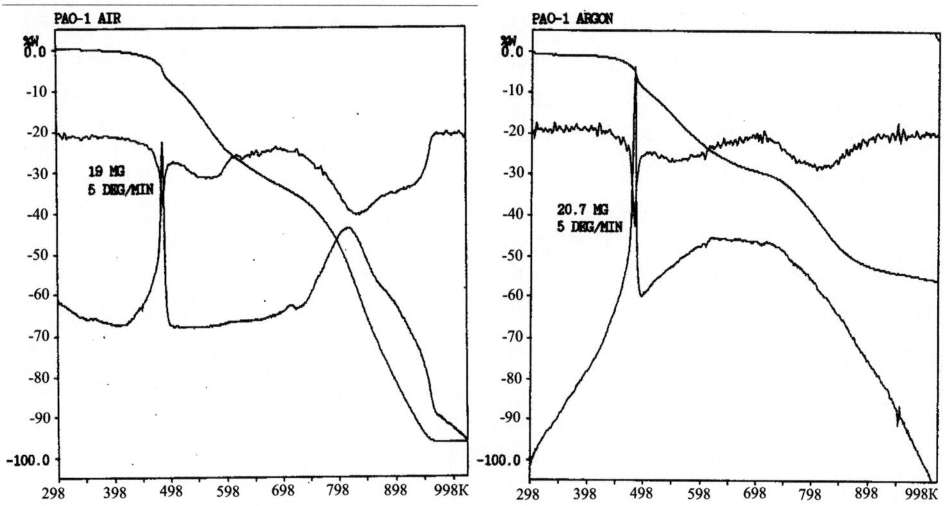

Fig. 4. The data of thermal analysis (TG, DTA, TMG) of PAO-I in argon and in the air

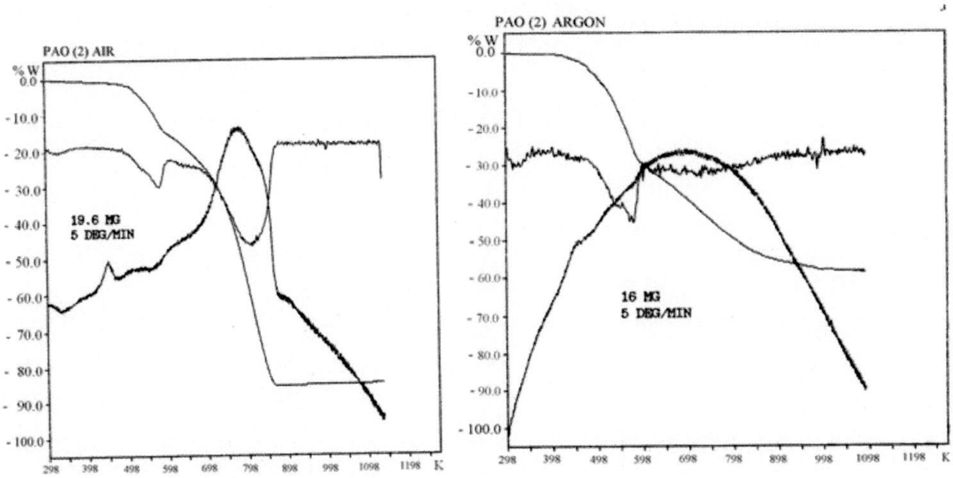

Fig. 5. The data of thermo analysis (TG, DTA, TMG) of PAO- V in argon and in the air

DSC analysis was used for glass temperature and melting point of PAO determination and presented values were in a good agreement with those obtained by thermomechanical and dielectrical methods. It is essential to note, that the regions of structural transitions predicted by increment modeling coincide within the limits of 5 % with DSC and thermomechanical analysis data.

The performed tests of dielectric strength using the high voltage generator BMW-30-01 at room temperature showed that PAO did not revealed electric conductivity even at high electric potential of about 3000 volt, i.e. PAO is a good dielectric.

Dielectric measurements of PAO-I demonstrated that this polymer reveals one dipole-segmental relaxation transition within the temperature range of 430 - 470K (Fig. 3), the nature of which is under research. Besides, PAO-I has rather high temperature of through conductivity around 470 K.

The thermal analysis showed that polyarylate oximates were of good thermal stability. It is evident from TGA curve for PAO-I that charcoal residue at 998 K in the air was about 5 % compared to 48 % in argon (Fig. 4). The charcoal residue for PAO -V containing card-like group was about 15 % in the air, whereas in argon this values came to 43 % (Fig. 5).

CONCLUSIONS

The results of the physical and chemical studies point out that polyarylate oximates obtained in this work revealed stability in aggressive media, good dielectric characteristics, and high thermal stability. These results enable us to recommend the use of synthesized polymers in manufacturing of industrial products for various purposes.

The performed earlier studies on modification of PBT by polyformal oximates with the purpose of improvement of its operational properties brought us to positive results[3]. Polyarylate oximates and polyformal oximates belong to a class of polyesters and, therefore, one should expect a positive effect from their use as modifying additives for polyakylene terephthalates. The preliminary experiments afforded positive results.

REFERENCES

[1] Vinogradova, S.V., Vasnev, V.A. *Polycondensation Processes and Polymers.* Moscow: Science Publ., 2000, p. 372.

[2] Vinogradova, S.V., Vasnev, V.A., Vygodsky Ya.S. Card-like Polyheteroarylenes. Synthesis, Properties and Diversity. *Uspekhi Khimii,* 1996. v. 65, p. 266.

[3] RF Patent. IPC C 08 L 67/02 2005. *Polymer composition.* Musaev, Yu.I., Mashukov, N.I., Musaeva, E.B., Mikitaev, M.A., Kvashin, V.A. RF Patent, № 2004107019; Date of application 09.03.94. Date of positive decision 16.03.05.

In: Polymers, Polymer Blends, Polymer Composites… ISBN 1-60021-168-2
Eds: A.K. Mikitaev, M.K. Ligidov et al., pp.207-212

Chapter 30

THERMOSTABLE POLYBUTYLENE TEREPHTHALATE (PBT) MODIFIED WITH POLYFORMAL OXIMATES (PFO)

M. A. Mikitaev[1], Yu. I. Musaev[2]*, E. B. Musaeva[2], V. A. Kvashin[2], R. B. Fotov[2], A. I. Eid[1] and Yu.V. Korshak[1]

[1]D.I. Mendeleev University of Chemical Technology of Russia, Miusskaya Pl., 9, Moscow 125047, Russia
[2]H.M.Berbekov Kabardino-Balkarian State University, Chernyshevsky Str., 173, Nalchik 360004, Russia.

ABSTRACT

The physical and chemical properties of polybutylene terephthalte (PBT) modified with polyformal oximates (PFO) on the basis of di-acetophenyloxid dioxime were investigated by thermogravimetric analysis (TGA), melt flow index (MFI), and differential scanning calorimetry (DSC). It was shown that the addition of 0,5-1% PFO by weight to PBT increased the initial temperature of thermal-oxidative degratation by 25-65°C, and it was possible to change the melt flow index (MFI) to the values convenient for processing.

Key words: Polyformal oximates; polyethers; antioxidants; thermal stability; polybutylene terephthalate.

INTRODUCTION

It is known that, the polybutylene terephthalate (PBT) is one of the perspective and universal thermoplastic polymers belonging to polyesters. It is produced industrially in wide scale and has different applications as a constructional material. The growing world wide

* Correspondence to: Jury I. Musaev, Kabardino-Balkarian State University, Nalchik, KBR, Russia, Chernishevsky 173, 360004.

production of PBT and of various products from it requires improving of its physical-chemical properties [1].

Earlier, we have synthesized polyformal oximates (PFO) on the basis of di-acetophenyloxid dioxime and methylene cloride. The aromatic polyether has possessed a new conjunction of chemical fragments in a polymeric chain due to a structure of the initial monomers [2].

The results of the physical and chemical studies revealed that polyformal oximates was stable against the influence of the aggressive medium, exhibited high thermal stability and good dielectric characteristics. Because of the chemical structure of PFO, it was of interest to study its possible use as a modifier in PBT blends with the purpose of increasing its thermal stability during the processing, improving of the operational qualities, and enlarging the exploitation resource of articles [3].

EXPERIMENTAL PART

Preparation of PBT Blends

Samples were obtained by extruding the mixtures of granulated PBT (grade B-305) and appropriate amount of 1 % solution of polyformal oximates in chloroform, which was initially dried up under vacuum at temperature 100°C within two hours. The mixture was then extruded at temperature 100°C at rotational speed of 50-70 rpm. The obtained polybutylene terephthalate samples contained 0,05-1% PFO-1.

Instrumental Methods

Physical properties of the prepared samples were investigated by various analytical methods. Thermal behavior was analyzed under air by TGA instrument "MOM" (Hungary), the heating rate of samples was kept at 5 °C/min, the weight of a sample was 25 mg. The melt flow index (MFI), which determines the processing method for thermoplastics, was measured by the standard method of GOST 11645-73. For the estimation of MFI values the instrument IIRT-M2 was used. Dielectric properties of the obtained samples were investigated by the method of dielectric losses. Electric measurements were carried out with the help of the bridge by applying alternating current of 10^3 Hz and digital readout R-5058 in a temperature interval from 20 to 250°C. The error in measurements of a loss-angle tangent did not exceed 5%.

RESULTS AND DISCUSSION

Results of our study showed that physical and chemical properties (thermo-stability, electric strength) of polybutylene terephthalate samples containing PFO as modifying additive in a wide temperature range considerably exceeded the properties of the known samples.

Figures 1 and 2 represent the thermal analysis data for PFO and PBT modified by PFO, as well as for non- stabilized PBT and PBT industrially stabilized.

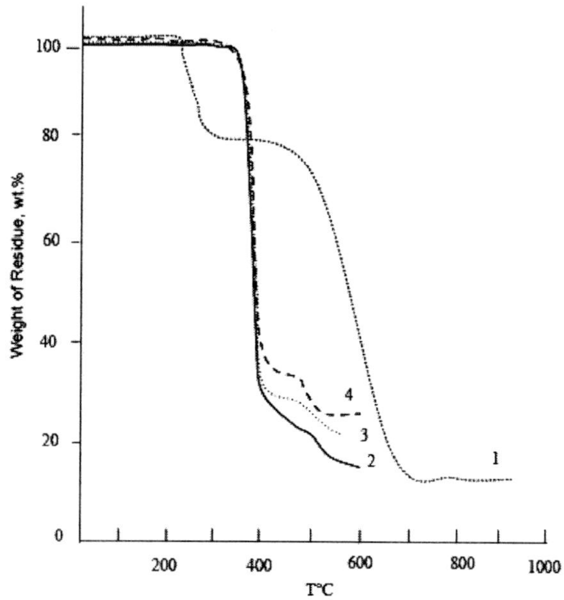

Fig. 1. TG analysis of various PBT/(PFO) blends with PFO content: (a) 0.5w % (curves 2, 3,) and (b) 1w % (curve 4). Curve 1is reffered to PFO

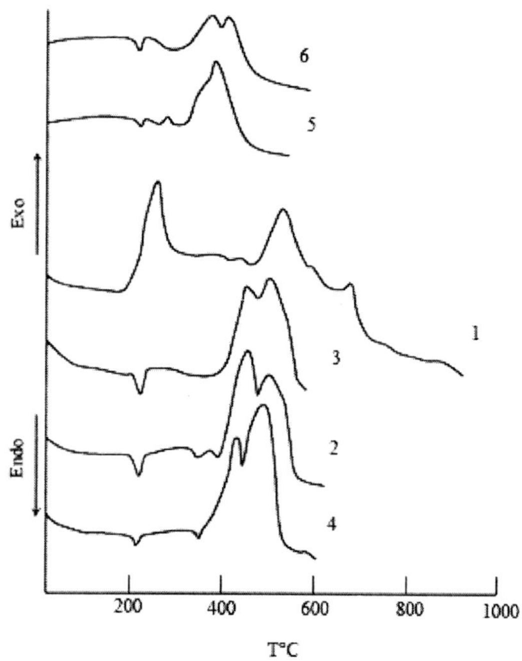

Fig. 2. Differential thermalal analysis (DTA) for non-stabilized (curve 5) and stabilized industrial PBT (curve 6) and for various PBT/PFO blends with PFO content: (a) 0.5w % (curves 2, 3); (b) 1w % (curve 4). Curve 1 is referred to PFO

It is evident from TGA curves (Fig.1) that the increase of the weight fraction of PFO-1 brings about the enlargement in mass fraction of the charcoal residue. Curve 3 is referred to a sample, which was exposed to preliminary thermal aging for 30 minutes at T = 250°C. The increase in the PFO-1 content displays a dual effect: simultaneously with the accelleration of thermal degradation the structural reorganization in samples takes place that is good agreement with the character of TG curves.

DTA data (Fig. 2 curves 5, 6) discloses that oxidation of non-stabilized and stabilized industrial PBT begins at 260 and 325°C, respectively. For PBT samples containing 0,5 wt % of PFO-1 (after 30 minute exposure at 250°C, curve 3) and 1 % wt. (curve 4), the oxidation process occurs at 350°C. The DTA curves of the samples shows active thermo-oxidative degradation with two peaks, the area and position of which depends on the PFO-1 contents (Fig.2).

DTA curves 5 and 6 (Fig. 2) show the peak corresponding to thermo-oxidation degradation with a bend at 462 - 466°C. The addition of the PFO-1 in the composite changes the character of DTA and DTG degradation and crosslinking becomes the basic process (curves 2-4). If PBT composite contains 0,5 % PFO-1 (without thermo-ageing) then no difference in character of DTA curves for PBT (non stabilized) and samples PBT + PFO- 1 in region of 460-470°C is observed. For a composite of PBT with 0,5 % PFO-1, which was held at 250°C for 30 minutes, the essential stabilizing effects of PFO-1 were found to be (the first peak corresponding to oxidation processes was reduced, while the second peak corresponding to crosslinking was increased).

In this respect, the best results were received for PBT samples with PFO-1 content of about 1 %. It can be seen from curve 4 (Fig. 2) that oxidation does not occur up to 390°C and crosslinking of composite happens in the range of 460-470°C. The first peak responsible for thermal-oxidative degradation dissapears almost completetely.

The melt flow index (MFI) was measured at 230°C and 2,16 kg for a PBT composite containing PFO-1 as a modifier and had a tendency to decrease up to 2,4 times (Fig. 3), depending on its content. Most likely, this is caused by increase in molecular weight of a polymer due to the chemical interaction between PBT and PFO-1 molecules at this temperature.

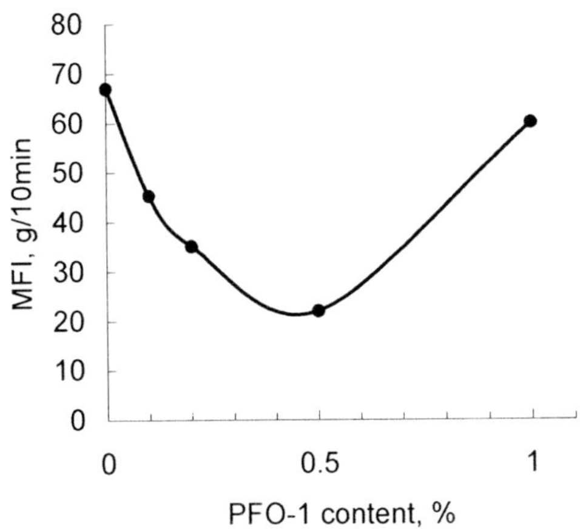

Fig. 3. A plot of melt flow index versus PFO-1 content in blends with PBT

A complex of positive effect resulting from the addition of PFO-1 in small quantities (~ up to 1. % wt) to PBT such as increasing of the coke residue and temperatures of the beginning thermal-oxidative degradation, and also the possibility to change MFI up to value convenient for PBT processing, PFO-1 can be used as a chemical modifier in a blends with PBT for the purpose of improving of its operational properties.

Addition of PFO-1 to a polybutylene terephthalate blends improves the dielectric properties, as well. This was proved by the results obtained from the dielectric analysis carried out on the blends of PBT +0,5% PFO-1 and PBT +1% PFO -1 (Fig. 4, Diagram 1).

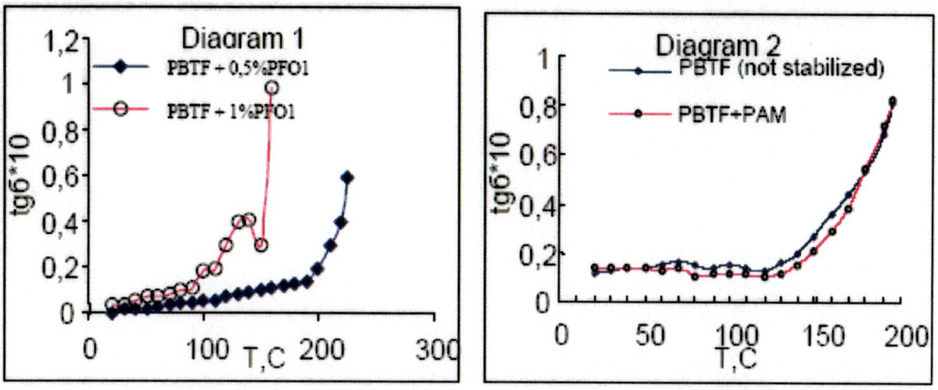

Fig. 4. A plot of loss-angle tangent versus temperature for PBT blends with PFO-1 (Diagram 1) and non-stabilized and stabilized with PAM PBT (Diagram 2)

The data on dielectric properties of industrial (non stabilized) PBT and PBT stabilized with PAM are submitted on the Diagram 2 (Fig. 4). It is evident from these data that molecular mobility in blends of PBT + 0,5% PFO -1 and PBT + 1% PFO-1 is a slightly higher than that for the industrial (not stabilized) PBT and compositions PBT/PAM in the range temperatures below glass temperature. Besides, a temperature of through conductivity is somewhat higher. Thus, for industrial non-stabilized PBT it is equal to 125°C, for PBT/polyazomethin – is 135°C, whereas for PBT +0,5% PFO-1 this temperature raises up to 190°C, and for PBT +1% PFO -1 mixture is 160°C.

CONCLUSIONS

A complex of positive effects of polyformal oximates (PFO) on PBT properties such as the increase of coke residue, of melting temperature, of the initiall temperatures of crosslinking and degradation was discovered. The change of a melt flow index up to the values convenient for PBT processing can be also reached. PFO in amount up to 1 % by weight can be used as a modifying additive to PBT for incraeasing its operational characteristics, e.g., the practical operational temperature and for broadening the temperature interval of processing.

REFERENCES

[1] Plachetta, Ch. *Polybutylene terephthalate* (PBT) / Kunststoffe. - 1995.85, N 10, with. 1588, 1590.

[2] Russian Federation Patent, [1] 2 223 977, MKE C 08 G 65/40 MPK C 08 G 65/40 *Polyformal and Polyether Formal and a Method of their Preparation.* / Musaev, Yu.I., Musaeva, E.B., Mikitaev, A.K., Hamukova, O.S. (Russian Federation),[1] 2 002 125 309/04; Applied 23.09.02. BI[1] 5, 2004.

[3] Russian Federation Patent, MKE C 08 L 67/022005. *A polymeric composition.* / Musaev, Yu.I., Mashukov, N.I., Musaeva, E.B., Mikitaev, M.A., Kvashin, V.A. (Russian Federation), [1] 2004107019; Applied 09.03.94. The positive decision from 16.03.05.

INDEX

A

B

C

D

E

J

K

L

M

N

O

P

Q

R

S

T